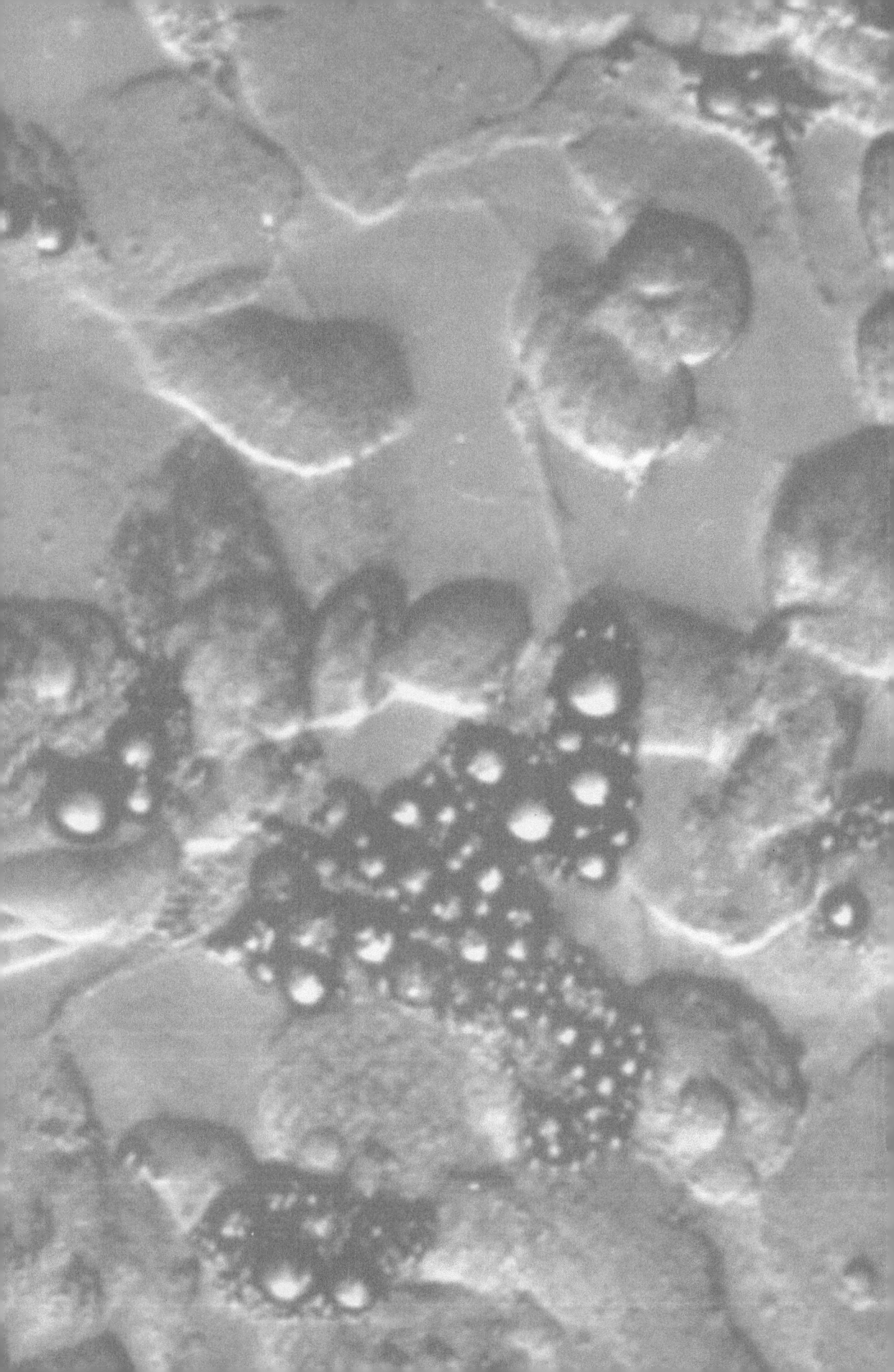

Effect of IL-10 and anti-TGF-beta antibodies on the morphology of bone marrow stromal cultures from

Interleukin-10

Jan E. DeVries and René de Waal Malefyt
R.G. Landes Co. 1995

MOLECULAR
BIOLOGY
INTELLIGENCE
UNIT

INTERLEUKIN-10

Jan E. de Vries

René de Waal Malefyt

DNAX Research Institute
Department of Human Immunology
Palo Alto, California, U.S.A.

SPRINGER-VERLAG BERLIN HEIDELBERG GMBH

MOLECULAR BIOLOGY INTELLIGENCE UNIT

INTERLEUKIN-10

International Copyright © 1995 Springer-Verlag Berlin Heidelberg
Originally published by Springer-Verlag in 1995
Softcover reprint of the hardcover 1st edition 1995

International ISBN 978-3-662-22040-5

Library of Congress Cataloging-in-Publication Data

Interleukin-10 / [edited by] Jan E. deVries, René de Waal Malefyt.
      p. cm.—(Molecular biology intelligence unit)
    Includes bibliographical references and index.
    ISBN 978-3-662-22040-5   ISBN 978-3-662-22038-2 (eBook)
    DOI 10.1007/978-3-662-22038-2

    1. Interleukin-10. I. de Vries, Jan Egbert. II. de Waal Malefyt, René. III. Series.
    [DNLM: 1. Interleukin-10. QW 568 I573 1994]
QR185.8.I56I584 1994
616.07'9—dc20
DNLM/DLC
for Library of Congress
    94-40332
    CIP

Publisher's Note

R.G. Landes Company publishes five book series: *Medical Intelligence Unit, Molecular Biology Intelligence Unit, Neuroscience Intelligence Unit, Tissue Engineering Intelligence Unit* and *Biotechnology Intelligence Unit*. The authors of our books are acknowledged leaders in their fields and the topics are unique. Almost without exception, no other similar books exist on these topics.

Our goal is to publish books in important and rapidly changing areas of medicine for sophisticated researchers and clinicians. To achieve this goal, we have accelerated our publishing program to conform to the fast pace in which information grows in biomedical science. Most of our books are published within 90 to 120 days of receipt of the manuscript. We would like to thank our readers for their continuing interest and welcome any comments or suggestions they may have for future books.

Deborah Muir Molsberry
Publications Director
R.G. Landes Company

CONTENTS

EDITORS

Jan E. de Vries
DNAX Research Institute
Department of Human Immunology
Palo Alto, California, U.S.A.
chapter 5

René de Waal Malefyt
DNAX Research Institute
Department of Human Immunology
Palo Alto, California, U.S.A.
chapters 3,5

CONTRIBUTORS

Jonathan B. Angel
Tufts University School
  of Medicine and the
  New England Medical Center
Boston, Massachusetts, U.S.A.
chapter 17

Jacques Banchereau
Schering-Plough Laboratory
  for Immunological Research
France
chapter 4

Dan Berg
DNAX Research Institute
Department of Immunology
Palo Alto, California, U.S.A.
chapter 16

Francine Brière
Schering-Plough Laboratory
  for Immunological Research
France
chapter 4

Edgar Carvalho
Servico de Immunologia
Universidade Federal de Bahia
Brazil
chapter 10

Amy E. Chernoff
Tufts University School
  of Medicine and the
  New England Medical Center
Boston, Massachusetts, U.S.A.
chapter 17

Robert L. Coffman
DNAX Research Institute
Department of Immunology
Palo Alto, California, U.S.A.
chapter 14

David L. Cutler
Schering-Plough Research Institute
Kenilworth, New Jersey, U.S.A.
chapter 17

CONTRIBUTORS

Charles A. Dinarello
Tufts University School
   of Medicine and the
   New England Medical Center
Boston, Massachusetts, U.S.A.
chapter 17

Ellen C. Donaldson
Tufts University School
   of Medicine and the
   New England Medical Center
Boston, Massachusetts, U.S.A.
chapter 17

Carl G. Figdor
Netherlands Cancer Institute
Department of Immunonolgy
The Netherlands
chapter 5

Rachel A. Freiberg
DNAX Research Institute
Department of Immunology
Palo Alto, California, U.S.A.
chapter 13

Eric V. Granowitz
Tufts University School
   of Medicine and the
   New England Medical Center
Boston, Massachusetts, U.S.A.
chapter 17

Alice S.-Y. Ho
DNAX Research Institute
Department of Immunology
Palo Alto, California, U.S.A.
chapter 1

Jeffrey S. Kennedy
Tufts University School
   of Medicine and the
   New England Medical Center
Boston, Massachusetts, U.S.A.
chapter 17

Neeltje A. Kootstra
Laboratory of Experimental
   and Clinical Immunology
University of Amsterdam
The Netherlands
chapter 11

Ralf Kühn
Institute for Genetics
University of Cologne
Germany
chapter 16

Gerhard Lonneman
Tufts University School
   of Medicine and the
   New England Medical Center
Boston, Massachusetts, U.S.A.
chapter 17

Frank Miedema
Central Laboratory Blood
   Transfusion Services
The Netherlands
chapter 11

Robert L. Modlin
Division of Dermatology
University of California
Los Angeles, California, U.S.A.
chapter 9

CONTRIBUTORS

Kevin W. Moore
DNAX Research Institute
Department of Immunology
Palo Alto, California, U.S.A.
chapter 1

Werner Müller
Institute for Genetics
University of Cologne
Germany
chapter 16

Anne O'Garra
DNAX Research Institute
Department of Immunology
Palo Alto, California, U.S.A.
chapter 6

Scott F. Orencole
Tufts University School
    of Medicine and the
    New England Medical Center
Boston, Massachusetts, U.S.A.
chapter 17

Fiona Powrie
DNAX Research Institute
Department of Immunology
Palo Alto, California, U.S.A.
chapter 14

Elaine Radwanski
Schering-Plough Research
    Institute
Kenilworth, New Jersey, U.S.A.
chapter 17

Steven G. Reed
Infectious Diseases Research
    Institute
Seattle, Washington, U.S.A.
chapter 8

Donna Rennick
DNAX Research Institute
Department of Immunology
Palo Alto, California, U.S.A.
chapter 16

Maria-Grazia Roncarlo
DNAX Research Institute
Department of Human
    Immunology
Palo Alto, California, U.S.A.
chapter 12

Nora Sarvetnick
Department of
    Neuropharmacology
Scripts Research Institute
La Jolla, California, U.S.A.
chapter 15

Hanneke Schuitemaker
Central Laboratory Blood
    Transfusion Services
The Netherlands
chapter 11

Leland Shapiro
Tufts University School
    of Medicine and the
    New England Medical Center
Boston, Massachusetts, U.S.A.
chapter 17

Alan Sher
Immunology and Cell Biology
    Section
Laboratory of Parasitic Diseases
Bethesda, Maryland, U.S.A.
chapter 8

CONTRIBUTORS

Myung-Shik Lee
Department of
    Neuropharmacology
Scripts Research Institute
La Jolla, California, U.S.A.
chapter 15

Peter A. Sieling
National Institutes of Health
laboratory of Parasitic Diseases
NIAID
Bethesda, Maryland, U.S.A.
chapter 8

Paul P. Trotta
Schering-Plough Research
    Institute
Biotechnology and
    Biochemistry Department
Kenilworth, New Jersey, U.S.A.
chapter 2

Edouard Vannier
Tufts University School
    of Medicine and the
    New England Medical Center
Boston, Massachusetts, U.S.A.
chapter 17

Peter Van Vlasselaer
Activated Cell Therapy
Mountain View, California,
    U.S.A.
chapter 7

Hei-De Wen
Tufts University School
    of Medicine and the
    New England Medical Center
Boston, Massachusetts, U.S.A.
chapter 17

William T. Windsor
Schering-Plough Research
    Institute
Biotechnology and Biochemistry
    Department
Kenilworth, New Jersey, U.S.A.
chapter 2

Sheldon M. Wolff
Tufts University School
    of Medicine and the
    New England Medical Center
Boston, Massachusetts, U.S.A.
chapter 17

Jiangchun Xu-Amano
DNAX Research Institute
Department of Immunology
Palo Alto, California, U.S.A.
chapter 1

Hans Yssel
DNAX Research Institute
Department of Immunology
Palo Alto, California, U.S.A.
chapter 3

Xi-Xang Zhang
Tufts University School
    of Medicine and the
    New England Medical Center
Boston, Massachusetts, U.S.A.
chapter 17

Albert Zlotnik
DNAX Research Institute
Department of Immunology
Palo Alto, California, U.S.A.
chapter 13

MOLECULAR BIOLOGY OF INTERLEUKIN-10 AND ITS RECEPTOR

Kevin W. Moore, Alice S.-Y. Ho and Jiangchun Xu-Amano

INTRODUCTION

The cytokine interleukin-10 (IL-10) was initially described as cytokine synthesis inhibitory factor (CSIF), a product of mouse T helper 2 (Th2) clones which inhibited cytokine synthesis—and therefore the effector functions—of mouse T helper 1 (Th1) clones.[1-3] IL-10 potently suppresses activation of macrophages, inhibiting release of monokines and the ability to serve as accessory cells for stimulation of T cell and natural killer (NK) cell function. IL-10 also is a costimulatory molecule for proliferation and differentiation of B cells, and can serve as a cofactor for stimulating growth and differentiation of T cells, mouse thymocytes, and mouse myeloid cells.[4,5] Here we discuss studies of recombinant IL-10 cDNA and genomic clones, the viral homolog BCRF1 (vIL-10), and IL-10 receptors.

IL-10 CDNA CLONES AND PROTEIN

Mouse IL-10 (mIL-10) and human IL-10 (hIL-10) cDNAs were cloned from a mouse Th2 cell line stimulated with concanavalin A, and from an activated tetanus toxin-specific human T-cell clone, respectively.[2,3] The cDNA clones expressed cytokines which inhibited IFNγ production in mouse and human CSIF assays. hIL-10 was active on both mouse and human cells, but mIL-10 was not active on human cells.[3,6] Rat IL-10 cDNA and predicted amino acid sequences have also been reported.[7] The mIL-10 and hIL-10 cDNA clones shared 80% DNA sequence homology, the only major difference being insertion of a human *Alu* repetitive sequence element in the 3'-untranslated region of the hIL-10 cDNA. The mIL-10 and hIL-10 cDNA clones both specify 178 amino acid open reading frames

Interleukin-10, edited by Jan E. deVries and René de Waal Malefyt.

(ORF)—including hydrophobic signal peptides—which are 73% identical. At least one monoclonal antibody cross-reacts with mIL-10 and hIL-10.[8]

Recombinant hIL-10 is an ~18 kDa polypeptide which appears to lack carbohydrate, while both recombinant and "natural" mIL-10 are N-glycosylated at Asn-8, resulting in a mixture of 17, 19 and 21 kDa polypeptides.[2,9] mIL-10 and hIL-10 appear to be expressed as acid-sensitive noncovalent homodimers,[10,11] but it is not completely established whether the monomeric cytokine is biologically active. Both cytokines contain two intrachain disulfide bonds, mIL-10 having a fifth cysteine which remains unpaired.[11] Reduction of the disulfide bonds abolished the cytokine's activities. IL-10 was predicted to be a member of the four α-helix bundle family of cytokines,[12] and does contain about 60% α-helix.[11] Biologically active recombinant mIL-10 and hIL-10 have been expressed in COS7 cells, mouse myeloma cells, Chinese hamster ovary cells, a baculovirus expression system, and *Escherichia coli*, although the *E. coli* product requires refolding to regain its biological activity. A hydrophilic peptide of eight amino acids can be attached to the N-terminus of both mIL-10 and hIL-10 without impairing biological activity.[8,13]

THE IL-10 GENE AND ITS EXPRESSION

The mIL-10 gene is encoded by five exons spanning ~5.5 kb of genomic DNA, and the m,hIL-10 genes are located on mouse and human chromosome 1.[14] Evidence for a similar structure of the hIL-10 gene has been obtained (S. Mohan-Peterson and R. de Waal Malefyt, unpublished). Transcription of the IL-10 gene appears to generate primarily a ~1.4 kb (mIL-10) or ~2 kb (hIL-10) mRNA,[2,3] although a shorter ~1 kb message was reportedly expressed by a mouse T helper 2 (Th2) clone.[2] Multiple potential transcriptional regulatory sequences were identified in noncoding regions of the mIL-10 gene,[14] including a NFκB-like recognition sequence, an AP-1 binding site, and IFN-responsive elements. The collection of potential transcriptional regulatory sequences resembled that found in mouse and human IL-6 genes.[14,15]

IL-10's initial characterization was as a Th2-derived inhibitor of Th1 cell function,[1] but it is now known to be expressed by a variety of cell types, including mouse mast cell lines,[2] a mouse CD8[+] T-cell clone,[16] mouse and human CD4[+] Th2 clones, T cells, B lymphomas, activated B cells and monocytes/macrophages, keratinocytes, as well as human CD4[+] Th$_0$ and Th1, and CD8[+] clones, and EBV-transformed B cells.[1-3,17-29] Thus, as suggested by sequence analysis of the IL-10 gene,[14] the pattern of IL-10 expression is similar to that of IL-6. mIL-10 production induced in vivo by LPS was not inhibited by cyclosporin, but anti-CD3 stimulated IL-10 expression in vivo was,[30] suggesting that stimulation of T cell IL-10 production (anti-CD3) could be inhibited by cyclosporin, but monocyte IL-10 expression (LPS) could not. In contrast, IL-10 synthesis by a mouse Th2 clone was reportedly not inhibited by cyclosporin.[31] This apparent discrepancy might be due to differences in the T cells—naive (anti-CD3) vs. the differentiated Th2 clone—or to the nature of the accessory cells and costimu-latory signals involved in the respective systems.

VIRAL HOMOLOGS OF IL-10

Both the Epstein-Barr virus (EBV) and equine herpesvirus type 2 (EHV2) genomes contain viral homologs of IL-10.[2,3,32] This relationship is most pronounced in the mature protein coding sequences, where for example hIL-10 and the EBV homolog BCRF1 (vIL-10) are ~84% identical, but declines virtually to insignificance in the signal sequence and flanking untranslated regions. Of mIL-10, hIL-10, and vIL-10, hIL-10 and vIL-10 encode the most closely related amino acid sequences, while mIL-10 and hIL-10 are significantly more closely related at the DNA level (~80%). Most of the divergence between hIL-10 and vIL-10 is found in the N-terminal 20 amino acids of the mature proteins[3,32] (Fig. 1.1). Not surprisingly, most anti-hIL-10 antibodies cross-react with vIL-10.[4,24] There is no horse IL-10 sequence available for comparison, but the EHV2 IL-10 sequence, like vIL-10, differs most from the mouse, rat, and human proteins in the N-terminal region[32] (Fig. 1.1).

Recombinant vIL-10 is a 17 kDa nonglycosylated polypeptide which shares some, but not all of the activities of IL-10,[3,6] although its specific activity is 3-10-fold lower than the cellular cytokine.[4,33] vIL-10 exerts CSIF activity on both mouse and human cells[3,6,24,34,35] and is a macrophage deactivating factor, inhibiting macrophage-dependent stimulation of NK cell cytokine synthesis[6,34,36] and superoxide anion production.[37] vIL-10 shares all known costimulatory activities of hIL-10 on human B cells,[38,39] and like mIL-10, it enhanced viability of mouse B cells in culture.[40] vIL-10 also enhanced reactivation of EBV-specific and HLA-unrestricted killer cell activity in PBMC from EBV+ donors.[41]

While no comparison of this latter activity with hIL-10 was reported, it is reminiscent of mIL-10's ability to promote differentiation of mouse cytotoxic T cells,[42] and of the observed enhancement of lymphokine activated killer activity by hIL-10.[34]

However, a number of the activities of mIL-10 and hIL-10 on mouse cells were not shared by vIL-10. vIL-10 neither enhanced class II MHC expression on mouse B cells,[40] nor served as a costimulator of mouse thymocyte or mast cell proliferation.[3,19] We were unable to demonstrate antagonism of hIL-10 activity by vIL-10 in these assays,[4] and as noted below, vIL-10 does not detectably antagonize binding of IL-10 to recombinant IL-10R.[8,13] These particular activities of IL-10 have up to now been characterized only on mouse cells, and parallel effects on human cells have not yet been reported. However, the lack of vIL-10 activity in these assays suggested the existence of multiple IL-10 receptor/signal transduction systems, only a subset of which interact with vIL-10.

Fig. 1.1. Alignment of amino acid sequences of rat IL-10,[7] mIL-10,[2] hIL-10,[3] BCRF1/vIL-10[2,3,67] and the IL-10-like gene from equine herpesvirus type 2 (EHV).[32] Vertical bars show amino acid residues which are identical between pairs of sequences.

What was gained by the ancestor of EBV when it acquired an IL-10 gene? vIL-10 is expressed in the late phase of the lytic cycle,[43] and vIL-10s ability to inhibit initial activation of anti-viral T and NK cells would be advantageous during this period, when viral proteins and infectious virus are produced and fresh B cells are being infected. This notion is consistent with results of studies of vIL-10-deleted EBV in which cells infected with the mutant virus were unable to inhibit IFNγ production when cultured with autologous peripheral blood cells.[44] Later, after establishment of latent infection, the ability of vIL-10 (and presumably cellular IL-10 as well) to enhance anti-viral cytotoxic activity could promote a balance between latency and lytic infection.[41]

Transformation of human B cells by EBV is complex and involves multiple viral gene products,[45,46] but the activities of vIL-10 on human B cells suggest a potential contribution to this phenomenon. Through its B cell-stimulating activities, vIL-10 could enhance the numbers of B lymphocytes and their susceptibility to infection/transformation. Transformation of B cells by EBV induced their production of hIL-10, and the outgrowth of EBV-transformed cells in vitro was delayed by the presence of an antiserum which blocked hIL-10 (and presumably also vIL-10) activity.[28] Miyazaki et al[29] reported vIL-10 expression as early as four hours after viral infection, with hIL-10 transcription subsequently detected at 12-18 hours. Moreover, anti-sense vIL-10 (but not control vIL-10 sense, or hIL-10 antisense) oligonucleotides blocked B cell transformation and EBV-induced Ig secretion in these cultures. The inhibitory effects of anti-sense vIL-10 on transformation were reversed by addition of exogenous hIL-10 to the cultures. In contrast to these findings, vIL-10-deleted EBV was reportedly unimpaired in its ability to transform human B cells.[44] It is hoped that further studies may illuminate the reasons for the apparently conflicting results.

Nonetheless, taken together, these studies suggest possible functions for vIL-10 and infection-induced expression of hIL-10 in the life cycle of EBV, and imply that vIL-10 confers a number of adaptive advantages on EBV in its interaction with the host immune system. We speculate that acquisition of the vIL-10 gene by the ancestral virus might have been an important step in evolution of EBV from pathogen to a-for the most part-harmless viral parasite.

IL-10 RECEPTORS

As found for other cytokines, the biological effects of IL-10 are mediated by interaction with cell-surface receptors. Recently mouse and human receptors for IL-10 (mIL-10R; hIL-10R) were identified and characterized.[8,10,13] Using ^{125}I-labeled hIL-10 as a probe for IL-10R, Tan et al[10] detected specific binding of IL-10 to a variety of mouse and human hemopoietic cell lines, including cells known to respond to IL-10. Binding affinity was high, with Kd in the 50-250 pM range. Only a few hundred receptors per cell were detected, a minimal estimate since virtually all cell types examined were themselves able to produce IL-10, and likely only "empty" IL-10R could be detected. Cross-linking studies with ^{125}I-hIL-10 indicated a molecular size of 90-120 kDa for hIL-10R, although a number of ^{125}I-labeled bands were detected, suggesting proteolytic degradation or the presence of multiple polypeptides cross-linked to labelled IL-10. mIL-10 bound only to mIL-10R, but hIL-10 bound both mouse and human cells, as expected from the species-specificity of IL-10.

FLAG-epitope[47]-labelled IL-10 was used to isolate mIL-10R cDNA clones from mouse mast cell and macrophage lines,[13] cell types which both respond to m,hIL-10; only the latter responds to vIL-10. Despite the difference in the m,hIL-10/vIL-10 responses of the parent cells, the IL-10R encoded by the mast cell and macrophage-derived cDNA clones were identical—the cDNA clones differed only in their 3'-untranslated regions. A parallel expression cloning approach utilizing FLAG-hIL-10 yielded hIL-10R cDNA clones from a Burkitt lymphoma cell line.[8] The hIL-10R cDNA clones exhibited 70% DNA sequence homology to the mast cell-derived mIL-10R cDNA. The mIL-10R and hIL-10R cDNAs encode open reading frames of 576 and 578 amino acids, respectively. The predicted hIL-10R protein is 60% identical and, if chemically similar amino acids are considered, exhibits 73% homology to the mIL-10R protein sequence. Cross-linking studies using ^{35}S-methionine-labelled IL-10, mIL-10R, and hIL-10R revealed single 90-110 kDa proteins expressed by mouse and human cell lines as well as by IL-10R-transfected COS7 cells. These data were substantially similar to those of Tan et al,[10] with some discrepancies likely attributable to the different labelled ligands employed or to fewer protease inhibitors used in the latter study. hIL-10 bound to recombinant m,hIL-10R, and mIL-10 bound only to mIL-10R, as expected.[4,10]

The functions of recombinant IL-10R were studied by expressing the receptors in a mouse pro-B cell line (Ba/F3), which did not respond to IL-10 and did not express IL-10R.[8,13] mIL-10R bound ^{125}I-hIL-10 with high affinity (Kd ~70 pM); the corresponding value for hIL-10R was 200-250 pM. The mIL-10R- and hIL-10R-expressing Ba/F3 cell lines both exhibited a proliferative response to IL-10. Interestingly, despite comparable levels of IL-10R expression, the hIL-10R-expressing line was about 10-fold less sensitive to hIL-10 than the mIL-10R transfectant.[8] In addition to proliferation, mIL-10R[+] Ba/F3 transfectants newly expressed a number of cell surface antigens in response to IL-10. Induction of these antigens was not observed with hIL-10R[+] Ba/F3, suggesting that hIL-10R was only partially functional in this mouse cell line. Site-directed mutagenesis and deletion analysis of the cytoplasmic domain of mIL-10R has indicated that different regions of the cytoplasmic domain are responsible for these two biological responses (A. S.-Y. Ho, S. H.-Y. Wei, and K.W. Moore, unpublished), as found for other cytokine receptors.[48-50]

mIL-10R mRNA was detected in a number of hemopoietic cells and cell lines, including cells known to respond to mIL-10 (B cells, thymocytes, mast cells, and macrophage cell lines). Similarly, hIL-10R mRNA was found mainly in human hemopoietic cells and cell lines. Activation of a number of human T-cell clones with anti-CD3 Mab and phorbol ester resulted in down-regulation of IL-10R mRNA levels,[8] providing evidence for regulation of IL-10R expression in T cells. While it is unclear whether even the observed 8-10-fold lower mRNA levels—and presumably hIL-10R number—would significantly impair the ability of activated T cells to respond to IL-10, the results suggested that, in contrast to resting T cells,[51,52] loss of IL-10R expression by activated T cells could render the latter insensitive to inhibition by IL-10 except for indirect effects mediated by macrophages/monocyte co-stimulatory cells.

mIL-10R and hIL-10R exhibted no primary sequence homology with other genes, but structural analysis of the predicted protein sequences showed they are novel members of the interferon receptor (IFNR) family.[8,13] Like IFNR, the extracellular portion of mIL-10R may be structurally divided into two homologous segments of ~110 amino acids.[53] The first of these domains features two conserved tryptophans and a cysteine pair, while a second, unique disulfide loop is formed in the membrane-proximal domain.[53] However, hIL-10R lacks the second cysteine in the second domain.[8] This structural relationship is intriguing because IL-10 and IFNγ antagonize each others' functions and production in several systems,[4,54] yet have some similar effects, such as induction of FcγR expression.[55] These findings suggest the possibility of interaction between the signalling pathways of IL-10R and IFNR. Structures of the ligand-binding chains of IFNαβR and IFNγR have been determined.[56-58] Biological evidence has for some time supported the existence of an additional IFNR polypeptide(s) involved in signal transduction (see for example ref. 59), one of which has recently been identified.[60,61] In view of the demonstration of shared subunits among receptors for different cytokines,[62,63] it is possible that IL-10R could share a second receptor subunit with an IFNR, although neither mIFNγ nor mIFNα competed for mIL-10 binding to mIL-10R,[13] and up to a 100-fold excess of hIL-10 did not compete for binding to hIFNγR on a human monocyte line, U937 (C.-C. Chou and C. Lunn, personal communication). An additional possibility is that the IL-10 and IFN signal transduction pathways may share signalling molecules, or that the IL-10/IL-10R interaction could directly antagonize the IFNR signal transduction pathway, perhaps by interacting with or sequestering one or more of its component proteins. These suggestions are consistent with evidence that IL-10 and IFNγ both induce FcγR expression[55] and activate a transcription factor which interacts with IFN-response elements in the FcγRI gene.[64]

Current evidence supports the likelihood of further IL-10R complexity. vIL-10, like the cellular cytokine, inhibits macrophage activation and accessory cell function, and stimulates mouse and human B lymphocytes, but has only a subset of the activities of hIL-10 on mouse cells,[3,19,40] suggesting a possible difference between IL-10R on vIL-10-responsive and -nonresponsive cells. Moreover, we found that vIL-10 did not compete effectively for IL-10 binding to recombinant IL-10R,[8,13] and FLAG-vIL-10 did not bind well to recombinant m,hIL-10R (S. H.-Y. Wei, J. Xu-Amano, Y. Liu, and K. W. Moore, unpublished). However, vIL-10 stimulates proliferation of mIL-10R+ Ba/F3 transfectants, with a specific activity 3-10-fold lower than hIL-10, as found on normal cells (A. S.-Y. Ho and K.W. Moore, unpublished). This observation suggests the possible presence of limiting amounts of another IL-10R component provided in *trans* by the host cell. Interestingly, hIL-10R+ Ba/F3 transfectants are considerably (40-100 fold) *less* sensitive to vIL-10 than are the mIL-10R⁺ cells. It remains to be seen whether these findings are explained by several distinct

IL-10R or by multiple IL-10R subunits, as observed for other cytokine receptors[62,63] and interferon receptors (IFNR).[60,61]

CONCLUDING REMARKS

In the four years since its first description,[1] IL-10 has become recognized as a cytokine which plays a crucial role in several areas of the immune system, most notably regulation of cell-mediated immune effector functions and in B cell proliferation and differentiation. The near future should bring increased understanding of IL-10s immunomodulatory roles(s) in vivo in normal and disease situations. The complete structure of IL-10R should emerge, and the mechanisms by which binding of IL-10 to IL-10R transduces a signal to the cell should also become clear; we may expect this pathway to share some common features with the IFN signal transduction pathway[64-66] and also to vary depending on the phenotype of the responding cell.

REFERENCES

1. Fiorentino DF, Bond MW, Mosmann TR. Two types of mouse helper T cell. IV. Th2 clones secrete a factor that inhibits cytokine production by Th1 clones. J Exp Med 1989; 170:2081-95.

2. Moore KW, Vieira P, Fiorentino DF et al. Homology of cytokine synthesis inhibitory factor (IL-10) to the Epstein Barr Virus gene BCRFI. Science 1990; 248:1230-34.

3. Vieira P, de Waal-Malefyt R, Dang M-N et al. Isolation and expression of human cytokine synthesis inhibitory factor (CSIF/IL10) cDNA clones: homology to Epstein-Barr virus open reading frame BCRFI. Proc Natl Acad Sci USA 1991; 88:1172-76.

4. Moore KW, O'Garra A, de Waal Malefyt R et al. Interleukin-10. Ann Rev Immunol 1993; 11:165-90.

5. Ho AS-Y, Moore KW. Interleukin-10 and its receptor. Therapeutic Immunology 1994; in press.

6. Hsu D-H, de Waal Malefyt R, Fiorentino DF et al. Expression of IL-10 activity by Epstein-Barr Virus Protein BCRFI. Science 1990; 250:830-32.

7. Goodman RE, Oblak J, Bell RG. Synthesis

and characterization of rat interleukin-10 (IL-10) cDNA clones from the RNA of cultured OX8- OX22- thoracic duct T cells. Biochem Biophys Res Commun 1992; 189:1-7.

8. Liu Y, Wei SH-Y, Ho AS-Y et al. Expression cloning and characterization of a human interleukin-10 receptor. J Immunol 1994; 152:1821-29.

9. Mosmann TR, Schumacher J, Fiorentino DF et al. Isolation of monoclonal antibodies specific for IL4, IL5, IL6, and a new Th2-specific cytokine (IL-10), cytokine synthesis inhibitory factor, by using a solid phase radioimmunoadsorbent assay. J Immunol 1990; 145:2938-45.

10. Tan JC, Indelicato S, Narula SK et al. Characterization of interleukin-10 receptors on human and mouse cells. J Biol Chem 1993; 268:21053-59.

11. Windsor WT, Syto R, Tsarbopoulos A et al. Disulfide bond assignments and secondary structure analysis of human and murine interleukin 10. Biochemistry 1993; 32:8807-15.

12. Shanafelt AB, Miyajima A, Kitamura T et al. The amino-terminal helix of GM-CSF and IL-5 governs high-affinity binding to their receptors. EMBO J 1991; 10:4105-12.

13. Ho AS-Y, Liu Y, Khan TA et al. A receptor for interleukin-10 is related to interferon receptors. Proc Natl Acad Sci USA 1993; 90:11267-71.

14. Kim JM, Brannan CI, Copeland NG et al. Structure of the mouse interleukin-10 gene and chromosomal localization of the mouse and human genes. J Immunol 1992; 148:3618-23.

15. Tanabe O, Akira S, Kamiya T et al. Genomic structure of the murine IL-6 gene: high degree of conservation of potential regulatory sequences between mouse and human. J Immunol 1988; 141:3875-81.

16. Hisatsune T, Minai Y, Nishisima K-I et al. A suppressive lymphokine derived from Ts clone 13G2 is IL-10. Lymphokine Cytokine Res 1992; 11:87-93.

17. O'Garra A, Stapleton G, Dhar V et al. Production of cytokines by mouse B cells: B lymphomas and normal B cells produce interleukin 10. Int Immunol 1990; 2:821-32.

18. O'Garra A, Chang R, Go N et al. Ly-1 B (B-1) cells are the main source of B-cell-derived IL-10. Eur J Immunol 1992; 22:711-717.

19. MacNeil I, Suda T, Moore KW et al. IL-10: a novel cytokine growth cofactor for mature and immature T cells. J Immunol 1990; 145:4167-73.

20. Fiorentino DF, Zlotnik A, Mosmann TR et al. IL-10 inhibits cytokine production by activated macrophages. J Immunol 1991; 147:3815-22.

21. Lin TZ, Svetic A, Ganea D et al. Cytokines in NZB CD5+ B clones. Annals NY Acad Sci 1992; 651:581-83.

22. Enk AH, Katz SI. Identification and induction of keratinocyte-derived IL-10. J Immunol 1992; 149:92-5.

23. Yssel H, de Waal Malefyt R, Roncarolo M-G et al. Interleukin 10 is produced by subsets of human CD4+ T cell clones and peripheral blood T cells. J Immunol 1992; 149:2378-84.

24. de Waal Malefyt R, Abrams J, Bennett B et al. IL-10 inhibits cytokine synthesis by human monocytes: an autoregulatory role of IL-10 produced by monocytes. J Exp Med 1991; 174:1209-20.

25. Salgame P, Abrams JS, Clayberger C et al. Differing lymphokine profiles of functional subsets of human CD4 and CD8 T cell clones. Science 1991; 254:279-82.

26. Yamamura M, Uyemura K, Deans RJ et al. Defining protective responses to pathogens: cytokine profiles in leprosy lesions. Science 1991; 254:277-79.

27. Benjamin D, Knoblach TJ, Dayton MA. Human B-cell interleukin 10: B cell lines derived from patients with AIDS and Burkitt's lymphoma constitutively secrete large quantities of interleukin 10. Blood 1992; 80:1289-98.

28. Burdin N, Peronne C, Banchereau J et al. Epstein-Barr virus transformation induces B lymphocytes to produce human interleukin-10. J Exp Med 1993; 177:295-304.

29. Miyazaki I, Cheung RK, Dosch H-M. Viral interleukin 10 is critical for the induction of B cell growth transformation by Epstein-Barr virus. J Exp Med 1993; 178:439-47.

30. Durez P, Abramowicz D, Gerard C et al. In vivo induction of interleukin-10 by anti-CD3

monoclonal antibody or bacterial lipopolysaccharide: differential modulation by cyclosporin A. J Exp Med 1993; 177: 551-55.

31. Wang SC, Zeevi A, Jordan ML et al. FK506, rapamycin, and cyclosporine: effects on IL-4 and IL-10 mRNA levels in a T-helper 2 cell line. Transplant Proc 1991; 23:2920-22.

32. Rode H-J, Janssen W, Rosen-Wolff A et al. The genome of equine herpesvirus type 2 harbors an interleukin-10 (IL-10)-like gene. Virus Genes 1993; 7:111-16.

33. Moore KW, Rousset F, Banchereau J. Evolving principles in immunopathology: interleukin 10 and its relationship to Epstein-Barr virus protein BCRF1. Springer Semin Immunopathol 1991; 13:157-66.

34. Hsu D-H, Moore KW, Spits H. Differential effects of interleukin-4 and -10 on interleukin-2-induced interferon-γ synthesis and lymphokine-activated killer activity. Int Immunol 1992; 4:563-69.

35. de Waal Malefyt R, Haanen J, Spits H et al. IL-10 and viral IL-10 strongly reduce antigen-specific human T cell proliferation by diminishing the antigen-presenting capacity of monocytes via downregulation of class II MHC expression. J Exp Med 1991; 174:915-24.

36. Tripp CS, Wolf SE, Unanue ER. Interleukin 12 and tumor necrosis factor a are costimulators of interferon γ production by natural killer cells in severe combined immunodeficiency mice with listeriosis, and interleukin 10 is a physiologic antagonist. Proc Natl Acad Sci USA 1993; 90:3725-29.

37. Niiro H, Otsuka T, Abe M et al. Epstein-Barr virus BCRF1 gene product (viral interleukin 10) inhibits superoxide anion production by human monocytes. Lymphokine Cytokine Res 1992; 11:209-14.

38. Rousset F, Garcia E, Defrance T et al. IL-10 is a potent growth and differentiation factor for activated human B lymphocytes. Proc Natl Acad Sci USA 1992; 89:1890-93.

39. Defrance T, Vanbervliet B, Briere F et al. Interleukin 10 and transforming growth factor β cooperate to induce anti-CD40-activated naive human B cells to secrete immunoglobulin A. J Exp Med 1992; 175:671-82.

40. Go NF, Castle BE, Barrett R et al. Interleukin 10 (IL-10), a novel B cell stimulatory factor: unresponsiveness of X chromosome-linked immunodeficiency B cells. J Exp Med 1990; 172:1625-31.

41. Stewart JP, Rooney CM. The interleukin-10 homolog encoded by Epstein-Barr Virus enhances the reactivation of virus-specific cytotoxic T cell and HLA-unrestricted killer responses. Virology 1992; 191:73-82.

42. Chen W-F, Zlotnik A. Interleukin 10: A novel cytotoxic T cell differentiation factor. J Immunol 1991; 147:528-34.

43. Hudson GS, Bankier AT, Satchwell SC et al. The short unique region of the B95-8 Epstein-Barr virus genome. Virology 1985; 147:81-8.

44. Swaminathan S, Hesselton R, Sullivan J et al. Epstein-Barr virus recombinants with specifically mutated BCRF1 genes. J Virol 1993; 67:7406-13.

45. Thorley-Lawson DA. Immunological responses to Epstein-Barr virus infection and the pathogenesis of EBV-induced diseases. Biochim Biophys Acta 1988; 948:263-86.

46. Tosato G. The Epstein-Barr virus and the immune system. Adv Cancer Res 1987; 49:75-125.

47. Hopp TP, Prickett KS, Price VL et al. A short polypeptide marker sequence useful for recombinant protein identification and purification. Bio/Technology 1988; 6: 1204-10.

48. Sakamaki K, Miyajima I, Kitamura T et al. Critical cytoplasmic domains of the common β subunit of the human GM-CSF, IL-3, and IL-5 receptors for growth signal transduction and tyrosine phosphorylation. EMBO J 1992; 11:3541-49.

49. Sato N, Sakamaki K, Terada N et al. Signal transduction by the high-affinity GM-CSF receptor: two distinct cytoplasmic regions of the common β subunit responsible for different signaling. The EMBO J 1993; 12:4181-89.

50. Fukunaga R, Ishizaka-Ikeda E, Nagata S. Growth and differentiation signals mediated by different regions in the cytoplasmic domain of granulocyte colony-stimulating factor. Cell 1993; 74:1079-87.

51. de Waal Malefyt R, Yssel H, de Vries JE.

Direct effects of IL-10 on subsets of human CD4+ T cell clones and resting T cells. J Immunol 1993; 150:4754-65.

52. Taga K, Mostowski H, Tosato G. Human interleukin-10 can directly inhibit T-cell growth. Blood 1993; 81:2964-71.

53. Bazan JF. Structural design and molecular evolution of a cytokine receptor superfamily. Proc Natl Acad Sci USA 1990; 87:6934-38.

54. Chomarat P, Rissoan M-C, Banchereau J et al. Interferon gamma inhibits interleukin 10 production by monocytes. J Exp Med 1993; 177:523-27.

55. te Velde AA, de Waal Malefyt R, Huijbens RJF et al. IL-10 stimulates monocyte FcgR surface expression and cytotoxic activity: distinct regulation of ADCC by IFNγ, IL-4, and IL-10. J Immunol 1992; 149: 4048-52.

56. Aguet M, Dembic Z, Merlin G. Molecular cloning and expression of the human interferon-γ receptor. Cell 1988; 55:273-80.

57. Uze G, Lutfalla G, Gresser I. Genetic transfer of a functional human interferon-α receptor into mouse cells: cloning and expression of its cDNA. Cell 1990; 60:225-34.

58. Bazan JF. Shared architecture of hormone binding domains in type I and II interferon receptors. Cell 1990; 61:753-54.

59. Pestka S. The interferon receptors: an unfinished story. AIDS Res Hum Retroviruses 1992; 8:776-86.

60. Soh J, Donnelly RJ, Kotenko S et al. Identification and sequence of an accessory factor required for activation of the human interferon gamma receptor. Cell 1994; 76:793-802.

61. Hemmi S, Bohni R, Stark G et al. A novel member of the interferon receptor family complements functionality of the murine interferon-gamma receptor in human cells. Cell 1994; 76:803-10.

62. Miyajima A, Hara T, Kitamura T. Common subunits of cytokine receptors and the functional redundancy of cytokines. Trends Biochem Sci 1992; 17:378-82.

63. Gearing DP, Comeau MR, Friend DJ et al. The IL-6 signal transducer, gp130: an oncostatin M receptor and affinity converter for the LIF receptor. Science 1992; 255: 1434-37.

64. Larner AC, David M, Feldman GM et al. Tyrosine phosphorylation of DNA binding proteins by multiple cytokines. Science 1993; 261:1730-33.

65. Muller M, Briscoe J, Laxton C et al. The protein tyrosine kinase JAK1 complements defects in interferon-α/β and -γ signal transduction. Nature 1993; 366:129-35.

66. Watling D, Guschin D, Muller M et al. Complementation by the protein tyrosine kinase JAK2 of a mutant cell line defective in the interferon-g signal transduction pathway. Nature 1993; 366:166-70.

67. Baer R, Bankier AT, Biggin MD et al. DNA sequence and expression of the B95-8 Epstein-Barr virus genome. Nature 1984; 310:207-11.

CHAPTER 2

PHYSICOCHEMICAL AND STRUCTURAL PROPERTIES OF INTERLEUKIN-10

Paul P. Trotta and William T. Windsor

INTRODUCTION

Interleukin-10 (IL-10) was first described as a murine factor produced by T helper 2 (Th 2) clones that could inhibit the production of various cytokines from Th 1 clones and hence was originally referred to as cytokine synthesis inhibition factor.[1] This novel activity was the basis for the isolation of cDNA clones coding for both murine IL-10[2] (mIL-10) and human IL-10[3] (hIL-10). Characterization of recombinant mIL-10 (rmIL-10) and recombinant hIL-10 (rhIL-10) has indicated pleiotropic stimulatory and suppressive activities on both lymphoid and myeloid cells (reviewed by Rennick et al[4] and Moore et al[5]). Interestingly, both mIL-10 and hIL-10 exhibit a high percentage of nucleotide and amino acid sequence homology to an open reading frame (BCRF1) in the Epstein-Barr virus genome.[2] The product of the BCRF1 gene has been termed viral IL-10 (vIL-10) since BCRF1 apparently represents a virally transduced hIL-10 gene that expresses many of the biological properties of hIL-10.[6]

The pleiotropic biological properties of rhIL-10 suggests that it may be useful in various clinical applications, particularly as an anti-inflammatory agent. This review will focus on our current knowledge of the physicochemical and biochemical properties of rhIL-10, including appropriate comparisons with the corresponding murine and viral proteins.

PREDICTED PHYSICOCHEMICAL AND STRUCTURAL PROPERTIES

The gene sequence of IL-10 provides a basis for predicting some physicochemical and structural properties of the mature protein, as summarized in Table 2.1. HIL-10 and mIL-10 encode open reading frames (ORFs) of 178 amino acids, compared to the somewhat shorter ORF for vIL-10 (170 amino acids).[2,3] Based on the consensus properties of the hydrophobic signal sequence,[7] a mature protein sequence of 160 amino acids is predicted for both hIL-10 and mIL-10, compared to 147 amino acids for vIL-10. N-terminal sequencing and electrospray mass spectrometry confirmed both the start sequence and length of the predicted mature rhIL-10.[8] Interestingly, analysis of metabolically radiolabelled mIL-10 suggests that the actual mature polypeptide sequence is 157 amino acids in length.[5] The predicted molecular weight values of the polypeptide chain of hIL-10, mIL-10 and vIL-10 are 18,647 18,453 and 17,355 Da, respectively based on their amino acid compositions. However, as noted below, the quaternary structure of rhIL-10 as well as rmIL-10 is a dimer composed of two identical polypeptide chains.[5,9,10] The murine genomic sequence contains four intron sites located between residue pairs Phe^{34}/Gln^{35}, Lys^{54}/Gly^{55}, Cys^{105}/His^{106} and Lys^{127}/Leu^{128}.[11]

The percentage identity of the amino acid sequences of these three IL-10 variants is high, ranging from 73% (human vs. murine) to 90% (human vs. viral). The sequence identity among these three variants is significantly lower for the first *ca.* 21 amino acids at the amino terminus (Fig. 2.1). It is notable that the mutational data matrix alignment indicates that 107 residues are identically conserved among the three IL-10 variants. However, only five residues are identical among the three hydrophobic leader sequences (Fig. 2.1).

All three forms of IL-10 are predicted according to the algorithm of Garnier et al[12] to have a secondary structure composed predominantly of alpha helix (Table 2.1). The prediction of beta sheet for each of these cytokines is less than 20%. As shown in Fig. 2.2, the alpha helical regions of hIL-10 are predicted to constitute four independent segments. Inclusion of additional criteria for the prediction of helical and turn regions suggests there may be up to six alpha helices.[10] These predictions are consistent with the four alpha helix bundle motif exhibited by a number of cytokines.[13-19]

There are two potential sites for N-linked glycosylation in mIL-10, and one each in hIL-10 and vIL-10. All three proteins may exhibit a maximum of two intramolecular disulfide bonds based on their cysteine content. The predicted isoelectric point of vIL-10 is acidic, compared to the somewhat alkaline values for the isoelectric points of mIL-10 and hIL-10 (Table 2.1).

Table 2.1. Predicted physicochemical and structural properties of IL-10[a]

	Human	Murine	Viral
• Total amino acids:	160(178)[b]	157(178)[b]	147(170)[b]
• Molecular weight Da:	18,647	18,453	17,351
• % Identity			
Human vs. Murine:	73%		
Human vs. Viral:	90%		
• No. of cysteines/disulfide potential:	4/2	5/2	4/2
• Potential N-glycosylation sites:	Asn[115]	Asn[8]; Asn[113]	Asn[104]
• Isoelectric point:	7.8	7.8	5.9
• Alpha helix[c]	75%	69%	83%
• Beta sheet[c]	9%	17%	7%

a   Predicted properties are based on the mature amino acid sequence as predicted from the cDNAs.[2,3]
b   Numbers in parentheses include the signal sequence.
c   Prediction was made according to Garnier et al.[12]

PHYSICOCHEMICAL PROPERTIES

ELECTROPHORESIS

Sodium dodecyl sulfate-polyacrylamide gel electrophoresis (SDS-PAGE) of highly purified rhIL-10 indicated a single species with an apparent molecular weight of *ca.* 17 kDa (Fig. 2.3). This value is consistent with the predicted polypeptide molecular weight (Table 2.1) and has been reported by others.[3,9] Significant amounts of dimer were not present under non-reducing conditions, indicating that the rhIL-10 dimer is non-covalent, as further described below. It is notable that Chinese hamster ovary (CHO) and *E. coli*-derived rhIL-10 demonstrated virtually identical electrophoretic migrations, a result that supports a lack of significant carbohydrate content in the mammalian cell-derived product (see below). In contrast, mammalian cell-derived rmIL-10 contains three species with apparent molecular weight values of 17, 19 and 21 kDa.[2] The larger two species were shown to be related to glycosylation-induced heterogeneity.[2] Isoelectric focusing-PAGE of CHO-derived rhIL-10 showed that the isoelectric point

(pI) of the major component was 7.92.[10] This value is very close to the predicted value of 7.8 (Table 2.1).

CARBOHYDRATE ANALYSIS

Carbohydrate analysis by the Dionex system indicated a lack of significant amounts of either N- or O-linked carbohydrate in CHO-derived rhIL-10 (R. Kumarasamy, unpublished observations). Thus, the single potential N-linked glycosylation site in rhIL-10 is unoccupied, as was also observed for vIL-10.[6] In distinction, Cos-derived rmIL-10, which has two potential N-linked carbohydrate sites, was shown by reaction with N-glycanase to contain N-linked glycosylation.[2] Based on the high sequence identity between mIL-10, vIL-10 and hIL-10, these results strongly suggest that the first N-linked gycosylation site (Asn[8]) in rmIL-10 is occupied. Glycosylation of rmIL-10 is not required for biological activity.[5]

SULFHYDRYL AND DISULFIDE CONTENT

The disulfide content of rhIL-10 and rmIL-10 was determined by reaction with 2-nitro-5-thiosulfobenzoate in the presence

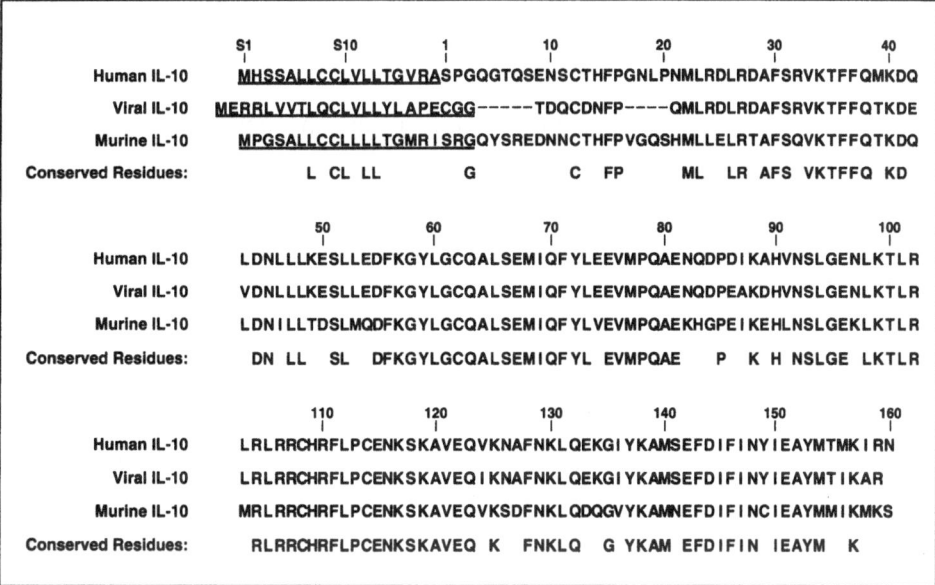

Fig. 2.1. Primary amino acid sequence alignment between human, viral and murine IL-10. Sequence alignment based on mutational data matrix.[22] N-terminal signal sequences are underlined. The identified conserved residues indicates identity between all three sequences.

of guanidine HC1.[8] RhIL-10 and rmIL-10 were found to have 2.15 ± 0.068 and 2.59 ± 0.25 moles of cystine/mole of protein, respectively. The number of free cysteine residues was quantitated by reaction with 4,4′-dithiodipyridine under denaturing conditions.[8] The results indicated 0.014 ± 0.007 and 1.03 ± 0.1 moles of sulfhydryl/mole of protein for rhIL-10 and rmIL-10, respectively. Thus, it was concluded that rhIL-10 and rmIL-10 each contained two intramolecular disulfide bonds, the maximum possible number. RmIL-10 apparently contains one cysteine as a free sulfhydryl group. These calculations employed an experimentally determined extinction coefficient for rhIL-10 of 8740 $M^{-1}cm^{-1}$ (or $\xi^{0.1\%} = 0.47\ mg^{-1}\ cm^2$) at 280 nm for protein concentration determination.

Disulfide Bond Assignment

Disulfide pairing was determined by proteolytic digestion with trypsin followed by identification of disulfide-containing peptides by either mass spectrometry or reversed-phase HPLC. Analysis of the data demonstrated unequivocally that for both rhIL-10 and rmIL-10 the first cysteine in the sequence was bonded to the third cysteine, and the second cysteine in the sequence was bonded to the fourth; i.e.,

Cys^{12} - Cys^{108} /Cys^{62} - Cys^{114} for rhIL-10 and Cys^9 - Cys^{105}/Cys^{59}- Cys^{111} for rmIL-10.[8] A partially reduced species of rhIL-10 contained a cleaved disulfide bond between Cys^{12} and $Cys,^{108}$ suggesting that this disulfide bond is solvent exposed in the native structure.[8]

STABILITY PROPERTIES

Full biological activity of rhIL-10 was retained for samples frozen at -20°C in phosphate-buffered saline, pH 7.0. In addition, lyophilization of rhIL-10 in ammonium bicarbonate, pH 8.5, followed by reconstitution in phosphate-buffered saline, pH 7.0, resulted in a fully active and structurally intact protein (W. Windsor, unpublished observations). Acidic treatment (below pH 5.5) of rhIL-10 and rmIL-10 led to biologically inactive material and dimer dissociation.[10]

STRUCTURAL PROPERTIES

SECONDARY STRUCTURE

The far ultraviolet (UV) circular dichroism (CD) spectrum of rhIL-10 was employed to estimate the percentages of secondary structural elements in rhIL-10 and rmIL-10. RhIL-10 purified to greater than 98% homogeneity from CHO cells displayed a spectrum with a shape and

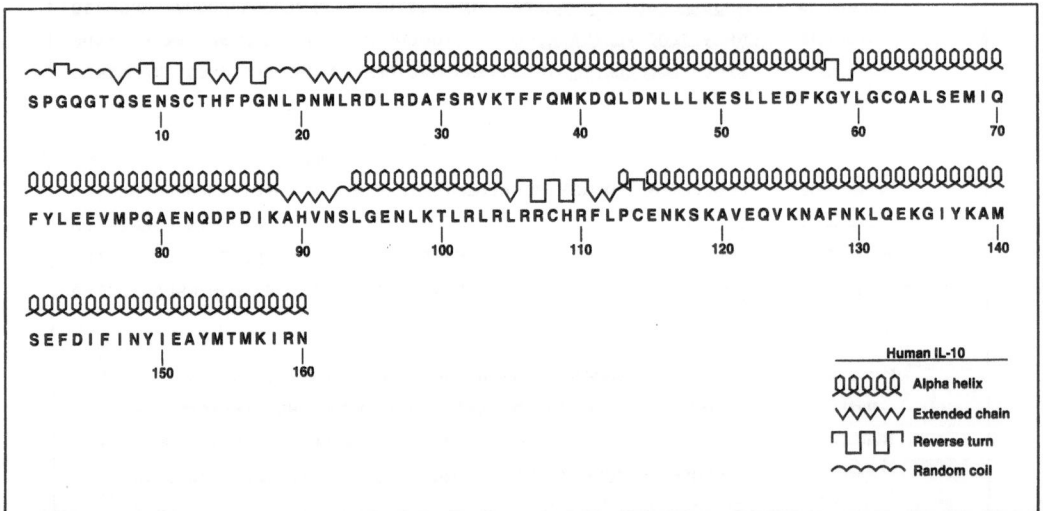

Fig. 2.2. Secondary structure prediction for mature human IL-10. The algoritham of Garnier et al[12] was applied utilizing a bias of greater than 50% for the α-helical content of hIL-10 based on estimates from far-UV CD spectra.

magnitude characteristic of the presence of a high degree of alpha helicity (Fig. 2.4A).[8] Analysis of the mean residue ellipticity spectrum indicated a content of 61% alpha helix, 7% beta sheet, 4% beta turn and 28% random structure. The measured alpha helical percentage was consistent with the high predicted value derived from an analysis of the amino acid sequence (Table 2.1). *E. coli*-derived rmIL-10 displayed a similar mean residue ellipticity spectrum, and hence apparently has a similar degree of alpha helicity, as predicted (Table 2.1). The corresponding values derived for the secondary structure of rmIL-10 were 60% alpha helix, 4% beta sheet, 8% beta turn and 28% random coil.

Reduction of rhIL-10 with dithiothreitol resulted in a significantly lower ellipticity between 200 and 230 nm (Fig. 2.4A).[8] This result suggests that reduction of the two disulfide bonds in rhIL-10 leads to a loss in alpha helical content and structural stability. Quantitative analysis of this spectrum indicated that the al-

pha helical content decreased to 53%. The random coil structure increased to 34%, and the amount of beta sheet (2%), and beta turn (11%) was essentially unchanged.

TERTIARY STRUCTURE

The high percentage of alpha helix content predicted from the primary sequence and calculated from the far-UV CD spectra suggests that both rhIL-10 and rmIL-10 may have a three-dimensional fold similar to the lymphokine and growth factor four alpha helical bundle superfamily.[13-19] The greater than 70% sequence identity between the human, murine and viral species strongly supports the presence of a similar tertiary fold.

Preliminary results from the analysis of quanidine HCL-induced unfolding of rhIL-10 measured by far-UV CD indicated that the protein undergoes a biphasic unfolding reaction.[10] Dissociation of the rhIL-10 dimer occurred below 1.5 M guanidine HCL. The second unfolding phase appeared to reflect the unfolding of

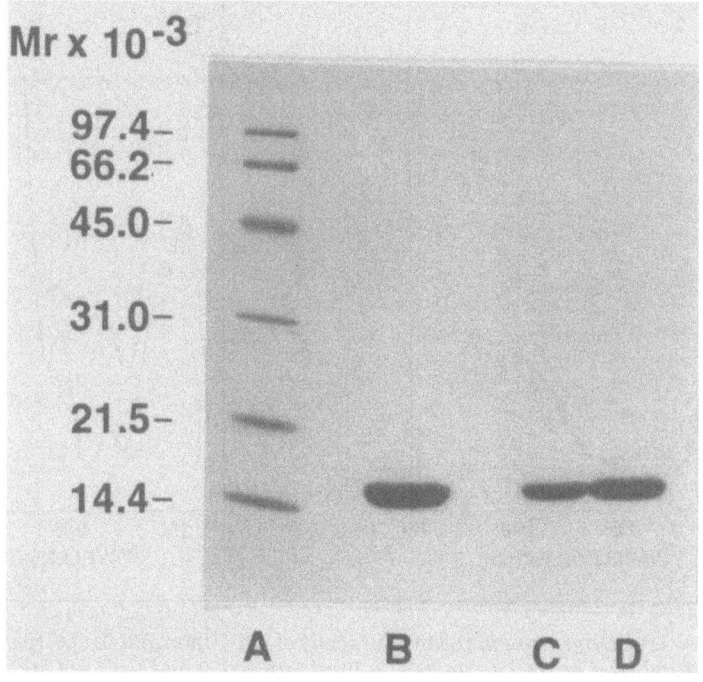

Fig. 2.3. SDS-PAGE of purified rhIL-10 under nonreducing conditions. Lane B, 10 μg CHO-derived rhIL-10; lanes C and D, 5 and 10 μg E. coli-derived rhIL-10, respectively; lane A, molecular weight markers. The stain was Coomassie blue R250.

the monomer, which has a [Gnd HCl]$^{1/2}$ of *ca.* 3.0 M. The relatively high stability properties of the monomer are likely to be due, at least in part, to its two disulfide bonds. The unfolding profile of the monomer is similar to that observed for other cytokines such as interleukin-4 and granulocyte-macrophage colony-stimulating factor.[20,21]

The near-UV circular dichroism spectrum was used to analyze the three dimensional environment of the aromatic amino acids. These spectra indicated significant differences in the optical rotation of rhIL-10 and rmIL-10 in this region (Fig. 2.4B).[8] This result signifies differences in the environment of the tyrosine residues despite the fact that four of the five tyrosines are conserved in position in the two proteins.

QUATERNARY STRUCTURE

The quaternary structure of rhIL-10 from CHO cells has been examined by chemical cross-linking, analytical ultracentrifugation and gel filtration. Although the apparent molecular weight of rhIL-10 in SDS-PAGE is consistent with the molecular weight of a single polypeptide chain, SDS-PAGE following chemical crosslinking with bis (sulfosuccinimidyl suberate) demonstrated species migrating with a molecular weight consistent with that of a rhIL 10 dimer.[10] Results from sedimentation equilibrium experiments also indicated that rhIL-10 exists primarily as a dimer at room temperature, pH 8.0, and at protein concentrations greater than 0.1mg/ml (P. Mui, unpublished observations). At a pH value of 3, rhIL-10 was observed to be predominantly monomeric in both cross-linking and sedimentation velocity experiments. These data indicated the existence of a non-covalent dimer that could be dissociated at low pH with 50% dimer dissociation occurring at *ca.* pH 4.5. Molecular shape calculations using sedimentation coefficients calculated from

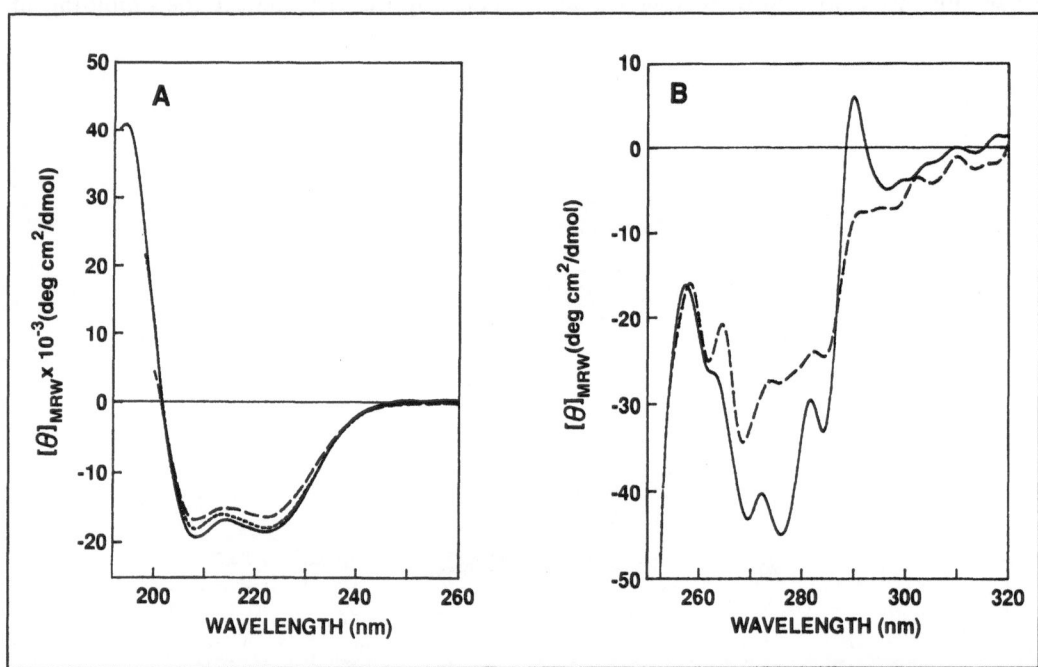

Fig. 2.4. Far-and near-UV CD mean residue ellipticity spectra of rhIL-10 and rmIL-10. (A) Far-UV CD: (–) rhIL-10 in 20 mM Tris and 10 mM NaCl, pH 8.1; (...) rmIL-10 in 20 mM Tris and 34 mM NaCl, pH 7.6; (–) rhIL-10 reduced with 1 mM dithiothreitol for 1 h at room temperature and purged with nitrogen (20 mM Tris and 10 mM NaCl, pH 8.1). The far-UV CD spectrum of the rhIL-10 sample at pH 7.6 was identical to that of the pH 8.1 sample. (B) Near-UV CD: (–) rhIL-10 in 20 mM Tris and 10 mM NaCl, pH 8.1; (–) rmIL-10 in 20 mM Tris and 0.15 M NaCl, pH 7.6. Reprinted with permission from Windsor, W. et al, Biochemistry. 1993; 32:8807-15.

sedimentation velocity experiments supported that the dimer was formed from an end-to-end association of monomers (P. Mui, unpublished observations).

Gel filtration experiments using a Superose 12 column were performed on rhIL-10 from CHO cells and *E. coli*-derived rmIL-10. The results yielded apparent molecular weight values of 37,000 and 33,000 Da, respectively (P. Mui, unpublished observations). These data as well as other gel filtration studies on rhIL-10 provide further evidence for a dimeric structure of IL-10.[9]

CRYSTALLIZATION

RhIL-10 has been crystallized from *ca.* 40% ammonium sulfate, 50mM HEPES, pH 7.2 (M.R. Walter and W.J. Cook, unpublished observations). A typical single crystal is shown in Fig. 2.5. The crystals were tetragonal, space-group $P4_3\, 2_1\, 2$, with unit cell axes of a=36.9, b=36.9 and c=222.3 Å. This highly asymmetric unit cell indicates an alignment of the long axis of rhIL-10 molecules along the long axis of the unit cell. A 3.5 Å map has been calculated and fit to a polyalanine backbone. Five alpha helices have been identified. This result is consistent with the prediction for at least four helical regions (Fig. 2.2) and with the general alpha helix bundle motif characteristic of several cytokines.[13-19]

FUTURE DIRECTIONS

A high priority of future studies will be to obtain a high-resolution crystal structure of rhIL-10. Heavy metal screening for use in solving the three-dimensional structure by multiple isomorphous replacement techniques has been initiated.

ACKNOWLEDGEMENT

We are grateful to Drs. Philip Mui, Nicholas J. Murgolo, Stephen H. Tindall, Ramasamy Kumarasamy, Mark R. Walter and William J. Cook, Mr. Paul Reichert, Ms. Rosalinda Syto and Mr. Vjay Khadse for communicating results prior to publication. Finally, we thank Ms. Barbara A. Sheldon for her excellent assistance in preparing the manuscript.

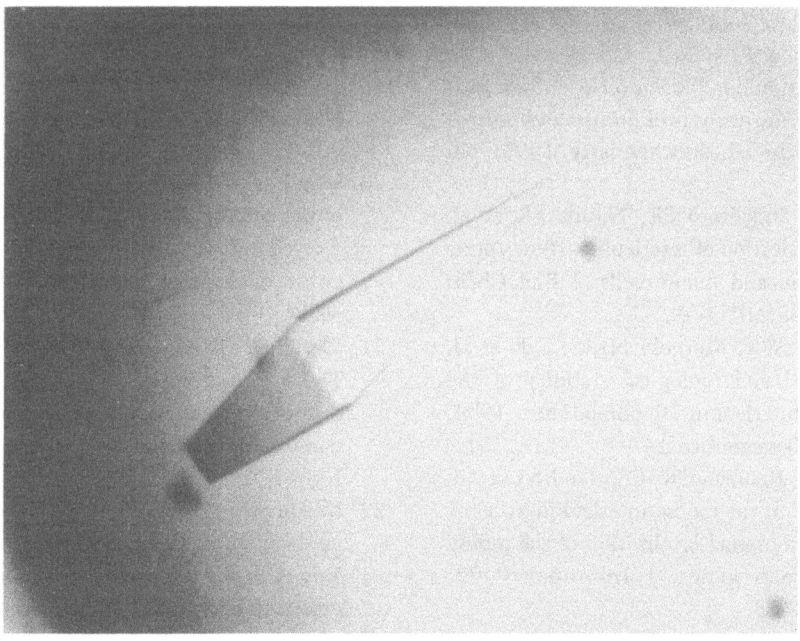

Fig. 2.5. Single crystal of rhIL-10 obtained by vapor-diffusion (see text; M. R. Walter and W. J. Cook, unpublished observations).

REFERENCES

1. Fiorentino DF, Bond MW, Mosmann TR. Two types of mouse helper T cell. IV. Th2 clones secrete a factor that inhibits cytokine production by Th1 clones. J Exp Med 1989; 170: 2081-95.

2. Moore KW, Vieira P, Fiorentino DF et al. Homology of cytokine synthesis inhibitory factor (IL-10) to the Epstein Barr virus gene BCRF1. Science 1990: 248:1230-34.

3. Vieira P, de Waal-Malefyt R, Dang M-N et al. Isolation and expression of human cytokine synthesis inhibitory factor (CSIF/IL10) cDNA clones; homology to Epstein-Barr virus open reading frame BCRF1. Proc Natl Acad Sci USA 1991; 88:1172-76.

4. Rennick D, Berg D and Holland G. Interleukin-10: an overview. Progress In Growth Factor Research 1992; 4:207-27.

5. Moore KW, O'Garra A, de Waal-Malefyt R et al. Interleukin-10. Annu Rev Immunol 1993; 11:165-90.

6. Hsu D-H, de Waal-Malefyt R, Fiorentino DF et al. Expression of IL-10 activity by Epstein-Barr virus protein BCRF1. Science 1990; 259:830-32

7. Perlman D, Halvorson MO. A putative signal peptidase recognition site and sequence in eukaryotic and prokaryotic signal peptides. J Mol Biol 1983; 167:391-409.

8. Windsor WT, Syto R, Tsarbopoulas A et al. Disulfide bond assignment and secondary structure analysis of human and murine Interleukin-10. Biochemistry 1993; 32: 8007-15.

9. Tan JC, Indelicato SR, Narula SK et al. Characterization of interleukin-10 receptors on human and mouse cells. J Biol Chem 1993; 268:21053-59.

10. Windsor WT, Murgolo NJ, Syto R et al. Structural and biological stability of the human Interleukin-10 homodimer. 1994, manuscript submitted.

11. Kim JM, Brannan CK, Copeland NG et al. Structure of the mouse interleukin-10 gene and chromosomal localization of the mouse and human genes. J Immunol 1992; 148:3618-23.

12. Garnier J, Osguthorpe DJ and Robson B. Analysis of the accuracy and implications of simple methods for predicting the secondary structure of globular proteins. J Mol Biol 1978; 120: 97-120.

13. McKay DB. Unraveling the structure of IL-2: Response. Science 1992; 257:412-13.

14. Ealick SE, Cook WJ, Vijay-Kumar S et al. Three-dimensional structure of recombinant human interferon-γ. Science 1991; 252: 698-702.

15. Walter MR, Cook WJ, Ealick SE et al. Three-dimensional structure of recombinant human granulocyte-macrophage colony-stimulating factor. J Mol Biol 1992; 224:1075-85.

16. Walter MR, Cook WJ, Zhao BG et al. Crystal structure of recombinant human interleukin-4. J Biol Chem 1992; 267: 20371-76.

17. Senda T, Shimazu T, Matsuda S et al. Three-dimensional crystal structure of recombinant murine interferon-β. EMBO J; 11:3193-3210.

18. Hill CP, Oslund TD and Eisenberg D. The structure of granulocyte-colony-stimulating factor and its relationship to other growth factors. Proc Natl Acad Sci USA 1993; 90:5167-71.

19. Milburn MV, Hassell AM, Lambert MH et al. A novel dimer configuration revealed by the crystal structure at 2.4 Å resolution of human interleukin-5. Nature. 1993; 363:172-76.

20. Windsor WT, Syto R, Le H et al. Analysis of the conformation and stability of E. coli-derived recombinant interleukin 4 by circular dichroism. Biochemistry 1991; 30:1259-64.

21. Wingfield P, Graber P, Moonen P et al. The conformation and stability of recombinant-derived granulocyte-macrophage colony stimulating factors. Eur J Biochem 1988; 173:65-72.

22. Needlemann SB, Wunsch CD. A general method applicable to the search for similarities in the amino acid sequences of two proteins. J Mol Biol 1970; 48:443-53.

CHAPTER 3

IL-10 AND HUMAN T CELLS

Hans Yssel and René de Waal Malefyt

INTRODUCTION

At least three subsets of CD4⁺ T cells with different cytokine production profiles can develop following stimulation with antigen in vivo. As was originally reported by Mosmann and collaborators, mouse T helper type 1 cells (Th1) are the primary producers of IL-2, IFN-γ and lymphotoxin (TNF-β), whereas they do not produce IL-4 and IL-5. T helper type 2 (Th2) cells on the other hand produce IL-4 and IL-5, but no IL-2, IFN-γ and TNF-β upon stimulation. Both subsets produce IL-3, GM-CSF and TNF-α (reviewed in ref. 1) In addition, a third subset has been described, designated Th0, in the mouse that is able to produce IL-2, IL-4, IL-5 and IFN-γ simultaneously.[2] A wealth of information has confirmed the existence of these different subsets in vivo and it is generally accepted that their induction is the consequence of chronic antigenic stimulation. For example, infection of mice with *Bordetella pertussis* has been shown to result in the development of a strong Th1 response, whereas antigen-specific T cells isolated from mice infected with *Nippostrongylus braziliensis* were of the Th2 type.[1]

Initially, attempts to classify CD4⁺ human T helper cells into subsets with different cytokine production profiles failed, since most of the T-cell clones, obtained from peripheral blood of healthy individuals, were of the Th0 type.[3,4] In addition, T-cell clones, specific for tetanus toxoid, alloantigens[3,5] or herpes simplex virus[6] were all able to produce IFN-γ and IL-4 simultaneously. However, later it was found that antigen-specific T-cell clones, established from the peripheral blood lymphocytes of patients with infectious diseases or of individuals with chronic allergic diseases, had a restricted cytokine production profile upon antigen-specific stimulation that was reminiscent of murine Th1 and Th2 type helper cells, respectively. Thus, Th1-like CD4⁺ T-cell clones could be established from the peripheral blood from patients with tuberculoid leprosy,[7] from patients infected with *Borrelia burgdorferi*,[8] or from donors reactive with PPD which is produced by *Mycobacterium tuberculosis*.[9] In contrast, allergen-specific T-cell clones isolated from the peripheral blood of

Interleukin-10, edited by Jan E. deVries and René de Waal Malefyt.
© 1995 R.G. Landes Company.

patients, allergic to the house dust mite allergen *Dermatophagoides pteronyssinus* (*Der pt*) or grass pollen (*Lolium perenne*),[10-12] from individuals with vernal conjunctivitis,[13] as well as from donors sensitized with *Toxocara canis* antigen[9] were mostly of the Th2 type. These findings in the human are consistent with those in the mouse, inasmuch as Th1 and Th2 subsets are detected most easily in situations of chronic antigenic stimulation.

Comparison between cytokine production profiles indicated that there are some differences between the two species. For example, IL-2 which is produced in mouse T-cell clones by Th1 cell only is produced by both human Th1 and Th2-like T-cell clones. Furthermore, as will be discussed in this chapter, the production of IL-10 in the human does not seem to be restricted to the Th2 type of T-cell clones.

PRODUCTION OF IL-10

Interleukin-10 was originally described in the mouse as a cytokine produced by Th2 T cells,[14] that had negative regulatory effects on the cytokine production and differentiation of Th1 cells.[15] In addition, IL-10 was found to be produced by Ly1 (CD5+) B cells, normal (CD20+, CD5-) B cells,[16,17] as well as by activated fetal thymocytes, as early as day 15 of gestation.[18] In addition, mouse keratinocytes can produce IL-10.[19,20] In man, IL-10 is produced by activated monocytes,[21] activated B cells,[22] B cells that are infected with Epstein Barr virus,[23] B cell lymphomas,[24] activated T cells and T-cell clones. IL-10 protein has also been detected in the serum of patients with non-Hodgkin's lymphoma[25] or multiple myeloma,[26] in ascites or biopsies of patients with ovarian and other intra-abdominal cancers[27] and finally IL-10 has been found to be produced by human carcinoma cell lines[28] and melanoma cells[29] in vitro. In this chapter we will focus on the production of IL-10 by T cells.

PRODUCTION OF IL-10 BY HUMAN PERIPHERAL BLOOD AND CORD BLOOD T CELLS

Human peripheral blood leukocytes, activated with the phorbol ester TPA and anti-CD3 monoclonal antibody (mAb) or with TPA and Ca^{2+} ionophore, produce IL-10. Monocytes are the major IL-10 producers in these cell samples, although peripheral blood T cells can produce significant amounts of IL-10, depending on the mode of activation (see below). Analysis of IL-10 production by peripheral blood T cells showed that IL-10 is both produced by CD4+, CD45RA+ "immunologically naive" T cells, as well as by CD4+, CD45RA- "memory" T cells. However, the amounts of IL-10 produced by the naive T-cell subset, upon activation with TPA and anti-CD3 mAb, were found to be 5-20 times lower than that of memory T cells and were ranging between 0.3-3 ng per 10^6 cells.[30] The finding that CD4+, CD45RA+ T cells present in peripheral blood were able to make low, but detectable levels of IL-10 following stimulation was confirmed using purified CD4+, CD45RA+ T cells isolated from cord blood (P. Schneider and H. Yssel, unpublished). This finding contrasts somewhat with an earlier report in which cytokine mRNA analysis by PCR showed that purified CD4+ cord blood T cells were only able to express transcripts for IL-2, but not for other cytokines, such as IFN-γ, IL-4 and GM-CSF.[31] However, no information about IL-10 was available at that time and therefore IL-10 production data were not included in this study. Furthermore, it should be noted that the expression of the CD45RA antigen on the surface of T cells may not delineate naive or memory cells per se and therefore the possibility that naive T cells do not produce IL-10 cannot be ruled out completely. CD8+ peripheral blood T cells, but not CD8+ cord blood T cells were found to be able to produce low levels of IL-10 following stimulation with TPA and anti-CD3 mAb (ref. 30, P. Schneider and H. Yssel, unpublished results).

IL-10 PRODUCTION BY HUMAN T CELL CLONES

In contrast to data reported for murine CD4+ T-cell clones, which indicated that IL-10 is exclusively produced by Th2 cells,[14] human CD4+ T-cell clones isolated from peripheral blood were found to produce IL-10, irrespective of their cytokine production profile. IL-10 could be detected in the culture supernatants of *Der pt*-specific Th2 T-cell clones, following antigen-specific activation, as well as after activation with TPA and anti-CD3 mAb. Similar stimulations of Th1-like human CD4+ T-cell clones, specific for *Borrelia burgdorferi*, *Yersinia enterocolitica* or *Mycobacterium leprae*, also resulted in the production of significant levels of IL-10.[30] Although the number of *Der pt*-specific Th2 cell clones included in this study was relatively small, the production of IL-10 by these cells, following antigen-specific activation, was in general higher than that of the Th1 T-cell clones. In a study from a different laboratory it was reported that 5/5 Th1 T-cell clones, specific for PPD, expressed IL-10 transcripts and produced IL-10 upon activation with TPA and anti-CD3 mAb. However, levels of IL-10 mRNA expressed by these Th1 T-cell clones were lower than those expressed by 6/6 *Toxocara canis*-specific Th2 T-cell clones.[32] Furthermore, support for the notion that Th1 cells can produce IL-10 was provided in a study in which the cytokine production profile of T-cell clones, specific for various mycobacterial antigens was analyzed.[33] As expected, these cells produced high levels of IFN-γ and low levels of IL-4, following stimulation. However, among these antigen-specific Th1 T-cell clones a large number of cells were found to produce low, but significant, levels of IL-10. Finally, in a recent study, we compared levels of IL-4, IL-5 and IL-10 produced by *Der pt*-specific T-cell clones that had been isolated from the peripheral blood of patients with atopic dermatitis. The production of IL-10 was found not to be correlated with the production of the Th2 cytokines IL-4 and IL-5 by these cells (C. Gutgesell and H. Yssel, unpublished results).

The reason for the species difference with respect to the production of IL-10 by mouse and human T cells is not clear and it is unknown at present whether these differences can be attributed to the different culture conditions used to expand human or mouse T cell clones. It is important to note however that in spite of the above mentioned studies, which failed to categorize IL-10 as a Th2 cytokine in human, a recent finding seems to be in some disagreement with this view. Whereas the production of IL-10 and that of Th2 cytokines produced by *Der pt*-specific T-cell clones, isolated from the peripheral blood of patients with atopic dermatitis was not correlated, as mentioned above, the production of IL-10 versus that of both IL-4 and IL-5 by *Der pt*-specific T-cell clones, isolated from skin lesions of the same patients showed a clear positive correlation (C. Gutgesell and H. Yssel, unpublished results). Also, the production of IFN-γ by the skin-derived T-cell clones was lower than that produced by the T-cell clones obtained from peripheral blood, whereas the production of IL-4, IL-5 and IL-10 was much higher. These findings not only indicate that the skin (and likely other organs where chronic allergic exposure takes place) may provide an unique environment for the development of T cells with a more pronounced Th2-like secretion profile, but also suggest that in such locations the production of IL-10 seems to be confined to the Th2 subset of CD4+ helper T cells.

KINETICS AND CONDITIONS OF IL-10 PRODUCTION

Kinetics studies have indicated that IL-10 is produced relatively late after activation. mRNA transcripts for IL-10 in stimulated peripheral blood T cells, as well as Th1 and Th2 clones are generally not detectable before 8 hours of activation and are still expressed after 24 hours. In contrast, mRNA transcripts for IL-2 and IL-4 are already maximally detectable after 4 hours of activation and are strongly reduced after 8 hours and 12 hours, respectively. The late transcription of IL-10 is reflected in the kinetics of IL-10 production by T-cell clones and peripheral blood T cells: IL-10 protein starts to be secreted at detectable levels at least 8-12 hours after activation by TPA and anti-CD3 mAbs. In contrast, the production of IL-4 and IFN-γ can already be measured 4 hours after stimulation.[12] Moreover, whereas the production of most cytokines, including IL-2, IL-4, IL-5 and TNF-α has declined after 24 hours of activation, the production of IL-10 by activated T-cell clones and peripheral blood T cells continues to increase at least until 48 hours after stimulation. It has been demonstrated that the production of IL-10 by activated monocytes also occurs considerably later than the production of other monokines.[21] In addition, IL-10 was found to down-regulate its own production by monocytes via an auto-regulatory feedback mechanism. A similar mechanism does not seem to be operational in human T cells (R. de Waal Malefyt, unpublished data). Nevertheless, the late kinetics of IL-10 production, together with its suppressive activities suggest a regulatory role for this cytokine in later phases of the immune response.

Although the levels of IL-10 production varied considerably among the individual T-cells clones, as mentioned above, the average levels of IL-10 production by Th2 T-cell clones following antigen-specific stimulation were higher than those produced by the Th1 clones.[30,32] Comparison of different modes of activation demonstrated that the production of IL-10 was maximally induced following stimulation with antigen or the combination of TPA and anti-CD3 mAb.[32] Interestingly, stimulation of T cells and T-cell clones with the combination of TPA and Ca²⁺ ionophore or ionomycin generally results in a poor production of IL-10, which is not due to insufficient activation of the cells, since this stimulation consistently induces high levels of other cytokines, particularly that of IFN-γ. Although little is known yet about the various transcription factors and DNA-binding proteins involved in the production of IL-10, it seems that the addition of Ca^{2+} ionophore has a negative regulatory effect on the production of this cytokine. Indeed, addition of Ca^{2+} ionophore to a T-cell clone, stimulated by TPA and anti-CD3 mAb resulted in the abrogation of IL-10 production by these cells, whereas still high levels of IL-4 and IFN-γ could be detected (S. Watanabe and H. Yssel, unpublished results) and suggests that the IL-10 signaling pathway in T-cell clones may be different from pathways leading to the production of other cytokines.

IL-10 PRODUCTION IN DISEASE

In addition to differences in cytokine production profiles, murine Th1 and Th2 cells were also found to differ in their functional properties: Th1 cells seem to be involved in cell-mediated delayed hypersensitivity reactions, whereas Th2 cells give optimal help in antibody production by B cells. It has to be noted however that antigen-specific T-cell responses are not always beneficial for the host. For example, BALB/c mice infected with *Leishmania major* develop a lethal Th2 responses, whereas, in contrast, C3H mice develop a protective Th1 response, following infection with the same organism.[34] BALB/c mice are protected against *L. major*-induced death by injection of neutralizing mAbs directed at IL-4, which prevents the generation of Th2 cells.[1]

Less information, however, is available about the consequences of the emergence of Th1 and Th2 responses in human diseases. Moreover, the role of IL-10 in the development of Th1 and Th2 subsets in human is not clear yet. Nevertheless, certain human diseases are characterized by the presence of monocytes or T cells, isolated from affected individuals, that produce varying amounts of IL-10. Analysis of cytokine transcripts extracted from leprosy skin biopsy specimen, using the polymerase chain reaction, showed a preponderance of Th1 cytokines in the lesions of the tuberculoid, resistant, form of the disease, whereas, in contrast, mRNA for IL-4, IL-5 and IL-10 predominated in the multi-bacillary form of the disease.[35] Similarly, elevated levels of IL-10 transcripts were present in bone marrow aspirates from patients with active visceral Leishmaniasis, a disease known as Kala-azar, whereas low levels of IL-10 transcription correlated with a resolution of the disease.[36] Together, these studies showed that resistance and susceptibility to certain diseases are correlated with relative high or low levels, respectively, of production of IL-10, similar to cytokine production profiles described in murine models of disseminated leishmanial disease.[1]

Finally, due to its many inhibitory effects, IL-10 seems to suppress antigen-specific T-cell responses in patients with lymphatic filariasis.[37] T cells, isolated from patients with an asymptomatic microfilaremic state which is characterized by the absence of antigen-specific T-cell responses, were found to produce high levels of IL-10, as compared to those from lymphatic filariasis patients with clinical signs of the disease. Since neutralizing anti-IL-10 mAb enhanced the in vitro antigen-specific T responses, it was speculated that the production of suppressive cytokines, such as IL-10 may result in situation of enhanced persistence of filarial parasites within humans which is characterized by the absence of clinical disease.

IL-10 has also been implicated in the induction and maintenance of transplantation tolerance and the prevention of graft-versus-host disease.[38] (see chapter 12)

DIRECT EFFECTS OF IL-10 ON HUMAN T CELLS

IL-10 INHIBITS THE PRODUCTION OF IL-2 BY T CELLS

Although IL-10 inhibits many T-cell functions by interference with antigen-presenting/accessory activities of APC, direct effects of IL-10 on survival, proliferation, cytokine production and chemotaxis by human T cells have been described. These direct effects of IL-10 on human T-cell function were substantiated by the cloning of a human IL-10 receptor and the demonstration of its expression on resting T cells, thymocytes and T-cell clones.[39]

Using a culture system that was devoid of monocytes, we demonstrated that IL-10 inhibited the proliferation of human T-cell clones, induced by anti-CD3 or anti-CD2 mAbs, crosslinked on murine L cells, transfected with FcγRII (CD32).[40] T-cell clones belonging to Th0, Th1 and Th2 subsets were equally affected by IL-10. This inhibition of T-cell proliferation was specific, since it could be reversed by a neutralizing anti-IL-10 mAb. Furthermore, mouse IL-10, which is not active on human cells, could not inhibit the proliferation of T-cell clones in this system, indicating that IL-10 did not exert its effect via the mouse L cells. However, the inhibitory effects of IL-10 were not complete and ranged from 20-70%. The mechanism of inhibition was found not to involve effects of IL-10 on costimulatory interactions between LFA-1 and ICAM-1, CD2 and LFA-3 or CD28/CTLA-4 and B7(CD80), respectively, since inhibition of T-cell proliferation was observed when L cells transfected with CD32 and ICAM-1, LFA-3 or CD80 were used. The observations that low amounts of IL-2 could reverse the inhibitory effects of IL-10 and that IL-10 did not inhibit the expression of IL-2R α or β chains by the T cells led to the finding that IL-10 inhibited the production of IL-2 by the T-cell clones. In fact, this effect was specific, since the production of IL-4, IL-5, IFN-γ and GM-CSF by activated T-cell clones was not affected by IL-10. The production of IL-2, but not that of IFN-γ and GM-CSF, was also inhibited when resting T cells isolated from peripheral blood were activated by anti-CD3 or anti-CD2 mAbs, crosslinked on CD32 L cells, in the presence of IL-10. However, co-expression of CD80 on the L cells was required for activation of the resting T cells.

Additional experiments on the mechanism of inhibition indicated that IL-10 also inhibited IL-2 production by human T-cell clones when these cells were activated by anti-CD3 mAbs and TPA, in the absence of L cells.[40] Furthermore, IL-10 inhibited the production of IL-2 by T-cell clones following activation by Ca²⁺ ionophore and

TPA, indicating that IL-10 does not affect immediate early signaling events, like protein phosphorylation, phosphoinositol breakdown, phospholipase C activation and Ca^{2+} fluxes. It was reported previously that IL-10 did not affect Ca^{2+} fluxes in T-cell clones, following activation by anti-CD3 mAbs.[41] The IL-10 mediated inhibition of IL-2 production occurred at the transcriptional level.[40] Currently, we are analyzing how IL-10 might affect the transcriptional activation of the IL-2 promoter.

Similar results, describing direct effects of IL-10 on human T cells, were reported by Taga et al[42] who demonstrated that IL-10 inhibited human T-cell proliferation and IL-2 production, following activation of resting T cells by immobilized anti-CD3 mAbs. IL-10 also inhibited the proliferation of human T cells, following activation by soluble anti-CD3 mAbs or PHA in the presence of monocytes. In this latter case, monocytes solely acted to cross-link the anti-CD3 mAb or fix PHA, thereby apparently not providing additional accessory signals that might be inhibited by IL-10. As such, the function of the monocytes in this system could be compared to the CD80+, CD32+ L cells used in our studies, described above.[40] Taken together, these results indicate that interaction of IL-10 with IL-10R on activated T cells in the absence of APC results in a specific inhibition of IL-2 production.

IL-10 INHIBITS APOPTOSIS

Activated human T cells and T-cell clones are dependent on IL-2 for growth and survival. Deprivation of IL-2 leads to rapid death by apoptosis, which involves loss of cell volume, chromatin condensation and DNA fragmentation.[43] Incubation of these T cells in the presence of IL-10 significantly reduces this apoptotic cell death.[44] This protective effect of IL-10 was independent of IL-2 and could not be reversed by anti-IL-2 mAbs or anti-IL-2R mAbs. Human T-cell clones cultured in the presence of IL-10 slightly increased the expression of IL-2R α chain (CD25) (R. de Waal Malefyt, unpublished). However, the

responsiveness of IL-2 deprived T cells cultured in the presence or absence of IL-10 to proliferate to IL-2 was indistinguishable.[44] These effects of IL-10 on cell survival and CD25 expression are not restricted to T cells. It has been shown that IL-10 enhances CD25 expression on human B cells activated by anti-CD40 mAbs, resulting in enhanced proliferation and immunoglobulin production.[45] In addition, it has been described that IL-10 enhances the survival of small splenic B cells, mast cells and macrophages in the mouse[46,47] and human monocytes (R. de Waal Malefyt, unpublished).

IL-10 MODULATES T-CELL CHEMOTAXIS

IL-10 has chemotactic activity on human T cells.[48] This chemotactic activity of IL-10 was subset specific and directed at CD8+ T cells, isolated from peripheral blood. No effects of IL-10 were found on the chemotactic activity towards CD4+ T cells, monocytes or neutrophil granulocytes. However, IL-10 did inhibit the IL-8 mediated chemotactic response of CD4+ T cells, but not that of CD8+ T cells. The reasons for this differential response of CD4+ and CD8+ T-cell subsets are not understood, but cannot be attributed to differential IL-10R expression, since both CD4+ and CD8+ T-cell clones expressed high levels of IL-10R.[39] These differential effects of IL-10 on T-cell chemotaxis could be important for the control of lymphocyte mediated immune reactions.

CONCLUDING REMARKS

The expression pattern and biological activities of IL-10 on human T cells support the hypothesis that IL-10 has a role in dampening immune responses. T-cell activation following encounter with antigen on antigen presenting cells leads to the rapid production of stimulatory cytokines and the initiation of late events in the immune response, such as proliferation and provision of B cell help. Once these events are set in motion, T cells start producing IL-10. This IL-10 will inhibit cytokine production, antigen presentation and activation of other T cells, both by its effects on the antigen presenting cells, as well as by its direct effects on T cells. As a result, the initiation of immune response is dampened and the effector phase of the immune response can clear the antigenic challenge. This mechanism can act in both Th1 as well as Th2 responses, since both human T helper subsets produce IL-10.

It is clear that disruption in the regulation of IL-10 production will have severe consequences for the individual which can lead either to immunosuppressed states, as observed in chronic parasitic infections when IL-10 is produced in excess, or to strong inflammatory reactions, as observed in mice with a targeted disruption of the IL-10 gene where no IL-10 is produced. These phenotypes provide strong evidence for the role of IL-10 as a negative regulator of immune responses under physiological conditions.

ACKNOWLEDGEMENT

DNA Research Institute of Molecular and Cellular Biology is supported by Schering-Plough Corporation.

REFERENCES

1. Mosmann TR, Coffman RL. Th1 and Th2 cells: different patterns of lymphokine secretion lead to different functional properties. Ann Rev Immunol 1989; 7:145-73.
2. Firestein GS, Roeder WD, Laxer JA et al. A new murine CD4+ T cell subset with an unrestricted cytokine profile. J Immunol 1989; 143:518-25.
3. Paliard X, de Waal Malefijt R, Yssel H et al. Simultaneous production of IL-2, IL-4 and IFN-g by activated human CD4+ and CD8+ T cell clones. J Immunol 1988; 141:849-55.
4. Maggi E, Del Prete G, Macchia D et al. Profiles of lymphokine activities and helper function for IgE in human T cell clones. Eur J Immunol 1988; 18:1045-50.
5. Bacchetta R, De Waal Malefijt R, Yssel H et al. Host-reactive CD4+ and CD8+ T cell clones from a human chimera produce IL-5, IL-2, IFN-γ and granulocyte/macrophage-colony-stimulating factor but not IL-4. J Immunol 1990; 144:902-08.
6. Yasukawa M, Inatsuki A, Horuchi T et al. Functional heterogenecity among herpes simplex virus-specific human CD4₊ T cells. J Immunol 1991; 146:1341-47.
7. Haanen JB, de Waal Malefijt R, Res PC et al. Selection of a human T helper type 1-like T cell subset by mycobacteria. J Exp Med 1991; 174:583-592.
8. Yssel H, Shanafelt M-C, Soderberg C et al. B. *burgdorfi* activates T cells to produce a selective pattern of lymphokines in Lyme arthritis. J Exp Med 1991; 174:593-601.
9. Del Prete GF, De Carli M, Mastromauro C et al. Purified protein derivative of *Mycobacterium tuberculosis* and excretory-secretory antigen (s) of *Toxocara canis* expand in vitro human T cells with stable and opposite (Type 1 helper or Type 2 helper) profile of cytokine production. J Clin Invest 1991; 88:346-50.
10. Wierenga EA, Snoek M, de Groot C et al. Evidence for compartmentalization of functional subsets of CD4+ T lymphocytes in atopic patients. J Immunol 1990; 144:4651-56.
11. Parronchi P, Macchia D, Piccini M-P et al. Allergen- and bacterial antigen-specific T-cell clones established form atopic donors show a different profile of cytokine production. Proc Natl Acad Sci USA 1991; 88:4538-42
12. Yssel H, Johnson KE, Schneider PV et al. T cell activation inducing epitopes of the house dust mite allergen Der *p* I. Induction of a restricted cytokine production profile of Der *p* I-specific T cell clones upon antigen-specific activation. J Immunol 1992; 148:738-45.
13. Maggi E, Biswas P, Del Prete G et al. Accumulation of Th-2-like helper T cells in the conjunctiva of patients with vernal conjunctivitis. J Immunol 1991; 146:1169-74.
14. Fiorentino DF, Bond MW, Mosmann TR. Two types of mouse T helper cells IV. Th2 clones secrete a factor that inhibits cytokine production by Th1 clones. J Exp Med 1989; 170:2081-95.
15. Fiorentino DF, Zlotnik A, Vieira P et al. IL-10 acts on the antigen-presenting cell to

inhibit cytokine production by Th1 cells. J Immunol 1991; 146:3444-51.

16. O'Garra A, Stapleton G, Dhar V et al. Production of cytokines by mouse B cells: B lymphomas and normal B cells produce interleukin 10. Int Immunol 1990; 2:821-32

17. O'Garra A, Chang R, Go N et al. Ly-1 B (B-1) cells are the main source of B cell derived IL-10. Eur J Immunol 1992; 149:92-5.

18. MacNeil IA, Suda T, Moore KW et al. IL-10: A novel growth cofactor for mature and immature T cells. J Immunol 1990; 145:4167-73.

19. Enk AH, Katz SI. Identification and induction of keratinocyte-derived IL-10. J Immunol 1992; 149:92-5.

20. Rivas JM, Ullrich SE. Systemic suppression of delayed-type hypersensitivity by supernatants from UV-irradiated keratinocytes. An essential role for keratinocyte-derived IL-10. J Immunol 1992; 149:3865-71.

21. de Waal Malefijt R, Abrams J, Bennett B et al. Interleukin 10 (IL-10) inhibits cytokine synthesis by human monocytes: an autoregulatory role of IL-10 produced by monocytes. J Exp Med 1991; 174:1209-20.

22. Vieira P, de Waal Malefyt R, Dang W et al. Isolation and expression of human cytokine synthesis inhibitory factor (CSIF/IL-10) cDNA clones: homology to Epstein Barr virus open reading frame BCRFI. Proc Natl Acad Sci USA 1991; 88:1172-76.

23. Burdin N, Peronne C, Banchereau J et al. Epstein-Barr virus transformation induces B lymphocytes to produce human interleukin 10. J Exp Med 1993; 177:295-304.

24. Benjamin D, Knobloch TJ, Dayton MA. Human B-cell interleukin-10: B cell lines derived from patients with acquired immunodeficiency syndrome and Burkitt's lymphoma constitutively secrete large quantities of interleukin-10. Blood 1992; 80:1289-98.

25. Blay JY, Burdin N, Rousset F et al. Serum interleukin-10 in non-Hodgkin's lymphoma: a prognostic factor. Blood 1993; 82:2169-74.

26. Merville P, Rousset F, Banchereau J et al. Serum interleukin-10 in early stage multiple myeloma. Lancet 1992; 340:1544-45

27. Gotlieb WH, Abrams JS, Watson JM et al. Presence of interleukin-10 (IL-10) in the ascites of patients with ovarian and other intra-abdominal cancers. Cytokine 1992; 4:385-90.

28. Gastl GA, Abrams JS, Nanus DM et al. Interleukin-10 production by human carcinoma cell lines and its relationship to interleukin-6 expression. Int J Cancer 1993; 55:96-101.

29. Chen Q, Daniel V, Maher DW et al. Production of IL-10 by melanoma cells: examination of its role in immunosuppression mediated by melanoma. Int J Cancer 1994; 56:755-61.

30. Yssel H, De Waal Malefyt R, Roncarolo M-G et al. IL-10 is produced by subsets of human CD4+ T cell clones and peripheral blood T cells. J Immunol 1992; 149:2378-84.

31. Ehlers S, Smith KA. Differentiation of T cell lymphokine gene expression: the in vitro acquisition of T cell memory. J. Exp. Med. 1991; 173:25-36

32. Del Prete G, De Carli M, Almerigogna F et al. Human IL-10 is produced by both type 1 helper (Th1) and type 2 helper (Th2) T cell clones and inhibits their antigen-specific proliferation and cytokine production. J Immunol 1993; 150:353-60.

33. Quayle AJ, Chomarat P, Miossec P et al. Rheumatoid inflammatory T cell clones express mostly Th1 but also Th2 and mixed (Th0-like) cytokine patterns. Scand J Immunol 1993; 38:75-82.

34. Coffman RL, Varkila K, Scott P et al. The role of cytokines in the differentiation of CD4+ subsets in vivo. Immunol Rev 1991; 123:189-207

35. Yamamura M, Uyemura K, Deans RJ et al. Defining protective responses to pathogens: cytokine profiles in leprosy lesions. Science 1992; 254:277-79.

36. Karp CL, El-Safi SH, Wynn TA et al. In vivo cytokine profiles in patients with kala azar. Marked elevation of both interleukin-10 and interferon-gamma. J Clin Invest 1993; 91:1644-48.

37. King CL, Mahanty S, Kumaraswami V et al. Cytokine control of parasite-specific

anergy in human lymphatic filariasis. Preferential induction of a regulatory T helper type 2 lymphocyte subset. J Clin Invest 1993; 92:1667-73.

38. Bacchetta R, Bigler M, Touraine J-L et al. High levels of interleukin 10 production in vivo are associated with tolerance in SCID patients transplanted with HLA-mismatched hematopoietic stem cells. J Exp Med 1994; 179:493-502.

39. Liu Y, Wei H-Y, Ho S-Y et al. Expression cloning and characterization of a human IL-10 receptor. J Immunol 1994; 152:1821-29.

40. de Waal Malefyt R, Yssel H, de Vries JE. Direct effects of IL-10 on subsets of human CD4+ T-cell clones and resting T cells: specific inhibition of IL-2 production and proliferation. J Immunol 1993; 150:4754-65.

41. de Waal Malefyt R, Haanen J, Spits H, et al. IL-10 and viral IL-10 strongly reduce antigen specific human T cell proliferation by diminishing the antigen presenting capacity of monocytes via downregulation of class II MHC expression. J Exp Med 1991; 174:915-24.

42. Taga K, Mostowski H, Tosato G. Human interleukin-10 can directly inhibit T-cell growth. Blood 1993; 81:2964-71.

43. Cohen JJ. Programmed Cell death in the immune system. Adv Immunol 1991; 50:55-87.

44. Taga K, Cherney B, Tosato G. IL-10 inhibits apoptotic cell death in human T cells starved of IL-2. Int Immunol 1993; 5:1599-1608.

45. Fluckiger AC, Garrone P, Durand I et al. Interleukin-10 (IL-10) upregulates functional high affinity IL-2 receptors on normal and leukemic B cells. J Exp Med 1993; 178:1473-81.

46. Go NF, Castle BE, Barrett R et al. Interleukin-10 (IL-10), a novel B cell stimulating factor: unresponsiveness of X-chromosome-linked immunodeficiency B cells. J Exp Med. 1990; 172:1625-31.

47. Thompson-Snipes L, Dhar V, Bond MW et al. Interleukin-10: a novel stimulatory factor for mast cells and their progenitors. J Exp Med 1991; 173:507-10.

48. Jinquan T, Larsen CG, Gesser B et al. Human IL-10 is a chemoattractant for CD8+ T lymphocytes and an inhibitor of IL-8-induced CD4+ T lymphocyte migration. J Immunol 1993; 151:4545-51.

CHAPTER 4

INTERLEUKIN-10
AND B LYMPHOCYTES

Francine Brière and Jacques Banchereau

INTRODUCTION

IL-10 was identified in 1989 as an activity produced by T helper cell subset 2 (Th2) inhibiting the synthesis of cytokines by Th1 cells.[1] IL-10 was also characterized independently as a product of murine CD5[+] B cell lymphomas able to support the survival of mast cell clones[2] and to enhance the proliferation of thymocytes.[3] As Th2 cells more specifically control humoral responses,[4] effects of IL-10 on B lymphocytes have been readily tested and demonstrated.[5,6] These effects were most particularly expected because Epstein-Barr virus (EBV), which transforms human B lymphocytes, carries a gene, BCRF1, which is 84% homologous at the protein level to human IL-10[7] and is called viral IL-10 (vIL-10). In this review, we discuss the biological effects of IL-10 in B cell physiopathology.

IL-10 PRODUCING CELLS

In the mouse, IL-10 is secreted by activated Th2 clones, mast cell lines, T cells, macrophages, keratinocytes,[8] as well as B lymphomas,[3] Ly-1[+](CD5[+]) and Ly-1[-] (CD5[-]) peritoneal B cells following stimulation with LPS.[2] Significant levels of IL-10 are produced by LPS-stimulated B cells from the peritoneal cavity which is enriched in Ly-1 (B-1), but not by activated splenic B cells where conventional B cells predominate.[9]

In the human, IL-10 is produced by T cells, monocytes as well as carcinoma and melanoma cell lines.[8] Furthermore, both CD5[+] and CD5[-] human B lymphocytes can release IL-10 following activation either through their antigen receptor or their CD40 antigen[10] (Burdin et al, submitted for publication). A dual ligation of CD40 antigen and antigen receptor results in the production of levels of IL-10 which are at least additive. Interestingly, addition of IL-4 to cultures of activated B cells results in decreased IL-10 secretion but increased IL-6 production

Interleukin-10, edited by Jan E. deVries and René de Waal Malefyt.

(Burdin et al, submitted for publication). Peripheral blood B lymphocytes and monocytes, but not T cells from patients with systemic lupus erythematous, rheumatoid arthritis and Sjögren syndrome spontaneously secrete IL-10.[11]

EBV transformed B cell lines also produce large amounts of hIL-10.[10,12,13] In contrast, viral IL-10 can be detected in the supernatant of only a minority of EBV transformed B cell lines but it seems to play a crucial role early during the transformation process by EBV, as anti-sense oligonucleotides for vIL-10 mRNA added at time of viral infection prevented subsequent B cell transformation.[13] However, this observation is challenged by the powerful transforming capacity of EBV mutants whose BCRF1 gene has been deleted.[14] Interestingly, B cells are induced to produce significant amounts of hIL-10 after EBV transformation but hIL-10 neutralizing antibodies do not prevent the establishment of EBV-transformed cell lines[10] though it slows their growth considerably. B lymphocytes infected with EBV display rare vIL-10 transcripts a few hours after infection,[13] while earlier studies had indicated vIL-10 to be a "late" viral gene expressed only during the lytic phase of virus replication.[15]

IL-10 IN BIOLOGICAL FLUIDS AND NATURAL ANTI-IL-10 ANTIBODIES

Human IL-10 has been detected in biological fluids under various pathological circumstances. In particular, hIL-10 was found in the serum of 47/101 patients with active non-Hodgkin's lymphomas while it was detected only in 3/52 patients with partial or complete response.[16] IL-10 occurrence in the serum of patients with intermediate or high grade lymphomas was associated with a poor survival. Recent studies have indicated that IL-10 is produced by lymphoma cells (Blay, personal communication) as also observed with AIDS lymphomas.[17] The ability of B lymphocytes to produce and respond to IL-10 (see paragraph 5) suggests that this cytokine may act directly as an autocrine growth factor for certain human B lymphoid malignancies. IL-10 has also been detected in the serum of patients suffering from early stage multiple myelomas particularly those with solitary myelomas.[18] IL-10 is not detectable in the serum of patients with advanced disease. Unlike IL-6, the presence of IL-10 is associated with a good prognosis.

Human IL-10 is also present in the biological fluids of patients suffering from infections. Very high levels of IL-10 have been measured in the serum of patients suffering from malaria[19] and in the cerebrospinal fluid from patients with *Neisseiria meningitidis* infections where in this latter case, IL-10 was found to be associated with a better prognosis (Menetrier-Caux et al, submitted for publication). Finally, the ascites of patients with ovarian and other intra-abdominal cancers also contain IL-10, which does not seem to be produced by the tumor cells.[20]

Autoantibodies specific for various cytokines have been detected in certain human serum samples.[21] In recent studies, anti-IL-10 autoantibodies have been detected at a low frequency as only in 6 out of 1000 tested serum scored positive (Menetrier-Caux et al, submitted for publication). Interestingly, the IgG anti-IL10 autoantibodies inhibited IL-10 induced proliferation of a mouse pro-B cell line (Ba/F3) transfected with the human IL-10 receptor.

IL-10 AND B-CELL SURVIVAL

Murine IL-10, like mIL-4 augments expression of class II MHC on small resting mouse splenic B cells and is a viability maintenance factor for these cells in culture.[5] Murine as well as viral IL-10 do not enhance MHC class II expression on B cells from X-linked immune deficient (xid) mice though both IL-10 sustain the viability of those cells.[5] This suggests the complexity of IL-10 receptors which is discussed elsewhere in this volume. Human IL-10 was also found to enhance the survival of splenic B lymphocytes during 15 days culture period, an effect associated with enhanced expression of bcl-2 protein.[22]

However, in our studies, hIL-10 did not modify expression of HLA class II antigens on human resting tonsillar B cells possibly because of their already high level of constitutive HLA class II expression, nor did it permit their increased short term survival.

In contrast, addition of IL-10 to cultures of chronic lymphocytic leukemia (CLL) was found to decrease viable cell recovery in all cases. Flow cytometric analysis, DNA gel electrophoresis and Giemsa staining, revealed that IL-10 induces B-CLL cells to die from apoptosis. B-CLL cells undergoing apoptosis in response to IL-10 show decreased bcl-2 protein levels. Addition of IL-2, IL-4, IFN-γ and anti-CD40 mAb prevents the IL-10 mediated apoptosis of B-CLL cells. The pro-apoptotic effect of IL-10 is specific for B-CLL as cells from non-Hodgkin's lymphomas and hairy cell leukemias do not undergo apoptosis after IL-10 treatment.[23] Whether a subpopulation of normal B lymphocytes is sensitive to the pro-apoptotic effects of IL-10 remains to be determined.

IL-10 INDUCES B-CELL PROLIFERATION

Human IL-10 plays an important role in the proliferation of human B cells all along their different maturation stages. Fetal bone marrow B-cell precursor (CD10⁺sIg) freshly isolated or precultured on stromal cells, proliferate when activated through their CD40 antigen in the presence of IL-3 and further addition of IL-10 and IL-7 potentiates this effect.[24]

On mature B cells, hIL-10 enhances DNA replication of B lymphocytes stimulated through their antigen receptors (surface Ig) either by insolubilized anti-IgM antibody or *Staphylococcus aureus* Cowan (SAC) particles.[6] The costimulation is of lower magnitude than that observed with either IL-2 or IL-4. In contrast, when B cells are stimulated by cross-linking of their CD40 antigen (with either anti-CD40 antibody and mouse L cells transfected with FcγRII/CD32 or mouse L cells transfected with CD40-ligand), hIL-10 markedly enhances B cell growth and is as effective as IL-4 over a 10-day period. hIL-10

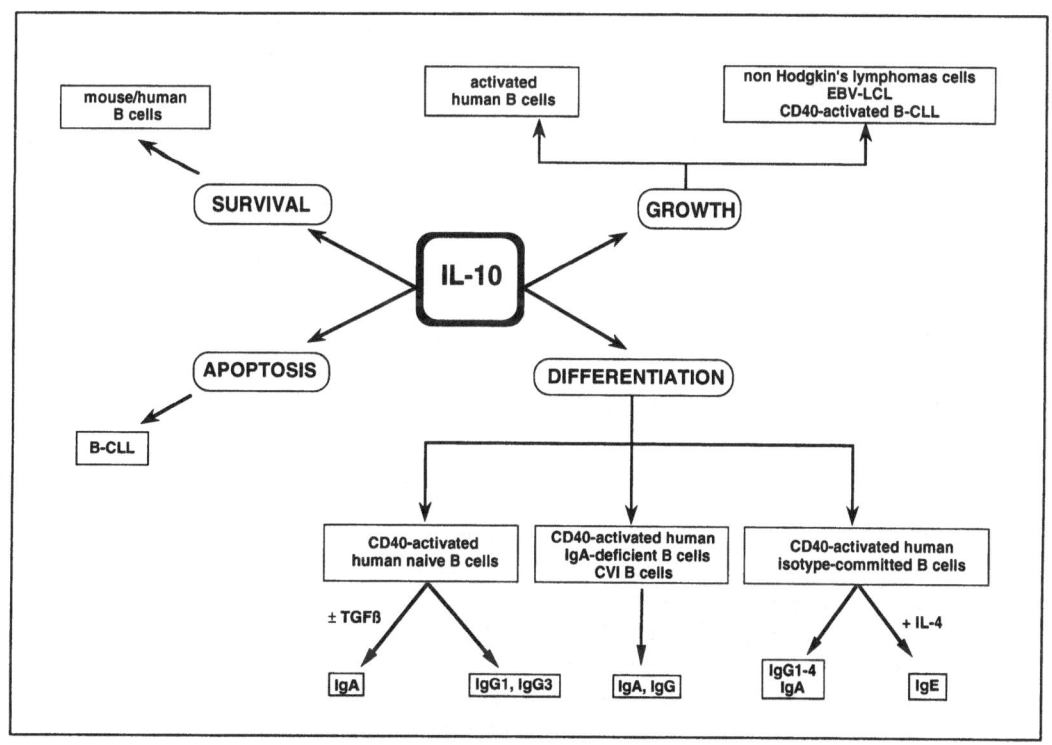

Fig. 4.1. Biological effects of IL-10 on B lymphocytes.

induces a 10-to-25-fold multiplication of B cells within 2 weeks. IL-4 and IL-10 are additive in those systems, their combination resulting in a 60-to-100 fold expansion of the number of viable B cells. Those signals are the most efficient tested so far, and during these 2 weeks surpasses even proliferation induced by EBV infection.[6] IL-10 also costimulates with IL-2 for the proliferation of B cells activated through CD40, an effect which is particularly striking as CD40-activated B cells respond poorly to IL-2. This synergy is likely to be due to the IL-10-induced increase of high affinity receptors on CD40-activated normal B lymphocytes or B-CLL cells which is associated to an increased Tac/CD25 expression.[25] On mature B cells, hIL-10 enhances EBV-induced B cell proliferation whereas IL-2 and IL-4 have no effect.[10] Interestingly, hIL-10 produced endogeneously upon EBV-transformation (see paragraph 2), plays a role in the growth of those B cells, as demonstrated by the addition of hIL-10 neutralizing antibodies which reduces considerably the EBV-induced B cell proliferation.[10]

Among mature B cells, subsets can be identified which differentially express specific surface molecules and which correspond to discrete stages of antigen-dependent B cell immunopoiesis.[26] In particular, naive B cells express sIgD, while sIgD- B cells are composed of two major subsets which include germinal center B cells and memory B cells.[27] IL-10 enhances equally well the DNA synthesis of the three different B cell subsets, be they stimulated through their antigen receptor or their CD40 antigen[28] (Liu and Brière, unpublished observations).

In contrast to human B lymphocytes, mIL-10 does not significantly enhance proliferation of LPS-activated mouse splenic B cells. Nevertheless, continuous administration of neutralizing anti-IL-10 antibodies to mice results in depletion of CD5+ B cells. However, this effect may be indirect through increased IFN-γ levels which inhibit CD5+ B cells growth.[29]

IL-10 INDUCES B-CELL DIFFERENTIATION

In vitro, hIL-10 induces SAC-preactivated human tonsillar B lymphocytes to produce high levels of IgM, IgG and IgA. Whereas IL-2 scores as the most potent cytokine for enhancing SAC-induced DNA synthesis, IL-10 appears to be the most efficient for induction of Ig synthesis by such cells.[6] The differentiation effect of IL-10 is particularly striking when B cells are stimulated through CD40. Unseparated CD40-activated B cells produce large amounts of IgM, IgA, IgG1, IgG2 and IgG3 but virtually no IgG4 nor IgE.[6,30] The IL-10 induced B cell differentiation does not require intense cross-linking of the CD40 antigen as soluble anti-CD40 mAbs and IL-10 are sufficient to induce Ig secretion.[31,32] Morphologic studies have shown that cells differentiate into plasmablasts full of intracytoplasmic Igs. Eventually, CD40-activated B cells stop proliferating after several days of culture in the presence of IL-10 possibly as a consequence of a terminal differentiation. Combination of IL-10 with IL-2 or IL-3 induces respectively SAC-preactivated B cells or T cell-preactivated B cells to terminally differentiate into plasma cells when secondary cultured onto peculiar microenvironments such as, rheumatoid arthritis synoviocytes (Dechanet et al, manuscript submitted) or bone marrow stroma (Merville et al, manuscript submitted). Noteworthy, IL-10 induces CD40-activated B-CLL cells to differentiate into plasma cells producing IgG, IgA and IgM.[25]

CD40-activated naive sIgD+ B cells isolated from tonsils, spleen, cord blood or adult blood were all found to respond to IL-10 by producing large quantities of IgM, as well as IgA, IgG1 and IgG3.[30,33] However, CD40 crosslinking is necessary for the effects on isotype switch as soluble CD40 antibody could induce human peripheral blood sIgD-memory B cells to secrete IgG in response to IL-10 but not sIgD+ B cells.[32] This observation suggests that IL-10 together with CD40 signaling may act as switch factor(s) for these IgG subclasses. In keeping with this, costimulation of those naive B cells with SAC and CD40 results in the production of large amounts of IgM without any other isotypes, thus excluding the possible contribution of contaminating post-switch B cells, which would fully differentiate in response to IL-10. Finally, combination of IL-10 and TGFß was found to induce naive CD40-activated B cells to produce large amounts of IgA.[33] In addition, IL-10 potentiates IL-4[31] or IL-13 (personal observation) dependent IgE synthesis induced by CD40-activated naive sIgD⁺ B cells. IL-10 has been shown to inhibit IL4-induced IgE production by unseparated blood mononuclear cells.[34] This is best explained by the strong inhibitory effect of IL-10 on monocyte dependent T-cell activation.[35]

IL-10 is also capable of enhancing Ig synthesis by isotype-committed B cells. In particular, in response to IL-10, CD40-activated memory B cells produce considerable amounts of IgG and IgA whereas germinal center B cells secrete much lower Igs (Liu and Brière, unpublished observations). Finally, CD40 triggering and IL-10 play an important role in Ag-specific antibody production by antigen-primed memory B cells.[32]

In vitro, IL-10 poorly affects Ig production by LPS-activated mouse splenic B cells. However, in vivo data suggest that IL-10 could also play a role in mouse B cell differentiation. Continuous administration of neutralizing anti-IL-10 antibodies to mice results in reduced serum IgM and IgA but increased serum IgG2a and IgG2b, depletion of peritoneal B cells, and a defective response to bacterial antigens such as phosphoryl choline and α1, 3-dextran.[29] Furthermore, IL-10 mRNA is upregulated 100-fold in spleen cells during the primary immune response of mice to goat anti-mouse IgD antibody which strongly activates humoral immunity.[36] However, IL-10 deficient mice do not display alteration in T and B lymphocyte development and have normal levels of the various serum isotypes.[37] Furthermore, when these animals develop chronic enterocolitis, they display increased serum IgA levels as a consequence of numerous plasma cells infiltrating the inflamed mucosa. The explanations for these discrepancies are not clear and may involve the well-recognized redundancy that exists within the cytokine system.[38] It is possible that a developing mouse embryo will compensate the loss of a critical cytokine gene by amplifying expression of a gene that codes for a protein of comparable function.

Neutralizing IL-10 may be beneficial in B-cell mediated autoimmune disease such as systemic lupus erythematosus. NZB/W F1 mice, which spontaneously develop an SLE-like syndrome with lethal immune complex dependent glomerulonephritis, display a delayed onset of autoimmunity when treated with anti-IL-10 and conversely injection of IL-10 accelerates the disease process.[39,40]

IL-10 IN B-CELL IMMUNODEFICIENCIES

IgA deficiency is the most common form of primary immunodeficiency in humans with a frequency of 1:300 in the Caucasian population, a third of whom being susceptible to infections. The molecular basis of this disorder is largely unknown and little information is available concerning the cytokines possibly involved in this disease. Peripheral blood lymphocytes from such IgA-deficient patients cultured in the CD40 system in the presence of SAC produce IgM and IgG but no IgA.[41] The addition of IL-10 to cultures enhances the production of IgM and IgG and, most strikingly, induces the production of high amounts of IgA. The IgA-inducing effect of IL-10 is direct as highly purified blood B cells from these patients can be induced to secrete IgA in response to CD40 engagement and IL-10. These patients do not appear to suffer from a major defect of IL-10 production as their blood mononuclear cells produced IL-10 following activation though levels were significantly reduced when compared to controls. More subtle alterations in the IL-10 cascade may be involved in the disease.

Common variable immunodeficiency (CVI) is an acquired disorder characterized by a failure of B cells to mature into Ig-secreting cells resulting in reduced Ig levels in serum and recurrent bacterial infections. It was therefore tested whether IL-10 could induce CVI B cells to secrete Igs. One study indicated that, in six out of seven patients tested, IL-10 could induce B cells to secrete IgG, IgA and IgM following activation with either anti-CD40 mAb or with SAC particles.[42] However, another study indicated a wide heterogeneity in the capacity of CD40-activated CVI B cells to produce different isotypes in response to IL-10, as either IgM only, or IgM and IgG or IgM, IgG and IgA could be secreted.[43] CD40 ligand on activated CVI-T cells was either not defective[42] or found at lower levels of mRNA expression.[44] Finally, in our own studies with 13 CVI patients, CD40-activated peripheral blood mononuclear cells were induced to secrete IgG and IgA in all cases in the presence of anti-CD40 mAb and IL-10 (Brière et al, submitted for publication). Thus, in most CVI cases, IL-10 appears able to overcome the inability of B cells to secrete IgG and IgA. Whether CVI is associated to defects in the IL-10 pathway remains to be determined.

CONCLUDING REMARKS

The effects of hIL-10 on B lymphocytes have been most convincingly demonstrated using human cells on which it displays strikingly different effects including apoptosis, proliferation and differentiation, according to their stage of maturation and/or activation. Such apparently contrasting effects are not specific to IL-10 since IL-4 can inhibit the proliferation of human B-cell precursors[45] but can enhance DNA synthesis of activated mature B cells.[46]

The activities of IL-10 on B lymphocytes suggest several clinical applications of either IL-10 agonists or IL-10 antagonists in diseases involving B cells. In particular, the proapoptotic effects of IL-10 on B-CLL cells calls for considering IL-10 or IL-10 agonists in the immunotherapy of advanced chronic lymphocytic leukemias. Administration of IL-10 to patients suffering from IgA deficiencies and common variable immunodeficiencies may permit to alleviate the block of B cell differentiation, thus possibly normalizing their deficient humoral responses. In contrast, IL-10 antagonists may prove useful in the management of systemic lupus erythematosus as well as non-Hodgkin's lymphomas. Such an antagonist may be (i) a soluble form of the receptor, (ii) a neutralizing antibody specific for IL-10, (iii) a chemical agent blocking the IL-10 cascade (not yet available) or (iv) interferons which have been shown to block the production of IL-10.[47]

Finally, studies aimed at identifying the respective intracellular pathways transducing IL-10 proapoptotic or proliferation/differentiation signals will hopefully permit to better understand the mechanisms underlying these different effector functions.

References

1. Fiorentino DF, Bond MW, Mosmann TR. Two types of mouse T helper cell. IV. Th2 clones secrete a factor that inhibits cytokine production by Th1 clones. J Exp Med 1989; 170:2081-95.

2. O'Garra A, Stapleton G, Dhar V et al. Production of cytokines by mouse B cells: B lymphomas and normal B cells produce Interleukin-10. Int Immunol 1990; 2: 821-32.

3. Suda T, O'Garra A, MacNeil I et al. Identification of a novel thymocyte growth promoting factor derived from B cell lymphomas. Cell Immunol 1990; 129:228-40.

4. Mosmann TR, Coffman RL. Heterogeneity of cytokine secretion patterns and functions of helper T cells. Adv Immunology 1989; 46:111-47.

5. Go NF, Castle B, Barrett R et al. Interleukin-10, a novel B cell stimulatory factor: Unresponsiveness of X chromosome-linked immunodeficiency B cells. J Exp Med 1990; 172:1625-31.

6. Rousset F, Garcia E, Defrance T et al. Interleukin-10 is a potent growth and differentiation factor for activated human B lymphocytes. Proc Natl Acad Sci USA 1992;

89:1890-93.

7. Hsu D-H, de Waal Malefyt R, Fiorentino DF et al. Expression of interleukin-10 activity by Epstein-Barr virus protein BCRF1. Science 1990; 250:830-832.

8. Moore KW, O'Garra A, de Waal Malefyt R et al. Interleukin-10. Annu. Rev Immunol 1993; 11:165-90.

9. O'Garra A, Chang R, Go N et al. Ly-1 B (B-1) cells are the main source of B cell-derived Interleukin-10. Eur J Immunol 1992; 22:711-17.

10. Burdin B, Péronne C, Banchereau J et al. Epstein-Barr virus-transformation induces B lymphocytes to produce human interleukin-10. J Exp Med 1993; 177:295-304.

11. Llorente L, Richaud-Patin Y, Wijdenes J et al. Spontaneous production of Interleukin-10 by B lymphocytes and monocytes in systemic lupus erythematosus. Eur. Cytokine Network 1993; 4:421-30.

12. Benjamin D, Knobloch TJ, Dayton MA et al. Human B-cell interleukin-10: B cell lines derived from patients with acquired immunodeficiency syndrome and Burkitt's lymphoma constitutively secrete large quantities of interleukin-10. Blood 1992; 80: 1289-98.

13. Miyazaki I, Cheung RK, Dosch H-M et al. Viral Interleukin-10 is critical for the induction of B cell growth transformation by Epstein-Barr virus. J Exp Med 1993; 178:439-47.

14. Swaminathan S, Hesselton R, Sullivan J et al. Epstein-Barr virus recombinants with specifically mutated BCRF1 genes. J Virol 1993; 67:7406-13.

15. Hudson GS, Bankier AT, Satchwell SC et al. The short unique region of the B95-8 Epstein-Barr virus genome. Virology 1985; 147:81-98.

16. Blay J-Y, Burdin N, Rousset F et al. Serum interleukin-10 in non-Hodgkin's lymphoma: A prognostic factor. Blood 1993; 82:2169-274.

17. Emilie D, Touitou R, Raphael M et al. In vivo production of interleukin-10 by malignant cells in AIDS lymphomas. Eur J Immunol 1992; 22:2937-42.

18. Merville P, Rousset F, Banchereau J et al. Serum Interleukin-10 in early stage multiple myeloma. The Lancet 1992; 340:1544-45.

19. Peyron F, Burdin N, Ringwald P et al. High levels of circulating IL-10 in human malaria. Clin & Exp Immunol 1994; 95:300-03.

20. Gotlieb WH, Abrams JS, Watson JM et al. Presence of Interleukin-10 (IL-10) in the ascites of patients with ovarian and other intra-abdominal cancers. Cytokine 1992; 4:385-90.

21. Bendtzen K, Svenson M, Jonsson V et al. Autoantibodies to cytokines: Friends or foes? Immunol Today 1990; 11:167-69.

22. Levy Y, Brouet J-C. Interleukin-10 prevents spontaneous death of germinal center B cells by induction of the bcl-2 protein. J Clin Invest 1994; 93:424-28.

23. Fluckiger A-C, Durand I, Banchereau J. Interleukin-10 induces apoptotic cell death of B-chronic lymphocytic leukemia cells. J Exp Med 1994; 179:91-99.

24. Saeland S, Duvert V, Moreau I et al. Human B cell precursors proliferate and express CD23 after CD40 ligation. J Exp Med 1993; 178:113-20.

25. Fluckiger A-C, Garrone P, Durand I et al. IL-10 upregulates functional high affinity IL-2 receptors on normal and leukemic B lymphocytes. J Exp Med 1993; 178:1473-81.

26. Banchereau J, Rousset F. Human B lymphocytes: phenotype, proliferation and differentiation. Adv Immunol 1992; 52:125-51.

27. Pascual V, Liu Y-J, Magalski A et al. Analysis of somatic mutation in B-cell subsets of human tonsil correlates with phenotypic differentiation from the naive to the memory B-cell compartment. J Exp Med 1994; 180: 329-39.

28. Lagresle C, Bella C, Defrance T. Phenotypic and functional heterogeneity of the IgD- B cell compartment: identification of two major tonsillar B cell subsets. Int Immunol 1993; 5:1259-68.

29. Ishida H, Hastings R, Kearney J et al. Continuous anti-IL-10 antibody administration depletes mice of Ly-1 B cells but not conventional B cells. J Exp Med 1992; 175:1213-20.

30. Brière F, Servet-Delprat C, Bridon J-M et

al. Human Interleukin-10 induces naive sIgD+ B cells to secrete IgG$_1$ and IgG$_3$. J Exp Med 1994; 179:757-62.

31. Garrone P, Galibert L, Rousset F et al. Regulatory effects of prostaglandin E2 on the growth and differentiation of human B lymphocytes activated through their CD40 antigen. J Immunol 1994; 152 42:82-90.

32. Nonoyama S, Hollenbaugh D, Aruffo A et al. B cell activation via CD40 is required for specific antibody production by antigen-stimulated human B cells. J Exp Med 1993; 178:1097-1102.

33. Defrance T, Vanbervliet B, Brière F et al. Interleukin-10 and Transforming Growth Factor ß cooperate to induce anti-CD40-activated naive human B cells to secrete Immunoglobulin A. J Exp Med 1992; 175:671-82.

34. Punnonen J, de Waal Malefyt R, Van Vlasselaer P et al. IL-10 and viral IL-10 prevent IL-4-induced IgE synthesis by inhibiting the accessory cell function of monocytes. J Immunol 1993; 151:1280-89.

35. de Waal Malefyt R, Abrams J, Bennett B et al. Interleukin-10 (IL-10) inhibits cytokine synthesis by human monocytes: an autoregulatory role of IL-10 produced by monocytes. J Exp Med 1991; 174:1209-20.

36. Svectic A, Finkelman FD, Jian YC et al. Cytokine gene expression after in vivo primary immunization with goat antibody to mouse IgD antibody. J Immunol 1991; 147:2391-97.

37. Kühn R, Löhler J, Rennick D et al. Interleukin-10-deficient mice develop chronic enterocolitis. Cell 1993; 75:263-74.

38. Finkelman FD, Holmes J, Katona IM et al. Lymphokine control of in vivo immunoglobulin isotype selection. Ann Rev Immunol 1990; 8:303-33.

39. Gérard C, Bruyns C, Marchant A et al. Interleukin-10 reduces the release of tumor necrosis factor and prevents lethality in experimental endotoxemia. J Exp Med 1993; 177:547-50.

40. Ishida H, Muchamuel T, Sakaguchi S et al. Continuous administration of anti-IL-10 antibodies delays onset of autoimmunity in NZB/W F1 mice. J Exp Med 1994; 179:305-10.

41. Brière F, Bridon J-M, Chevet D et al. Interleukin-10 induces B lymphocytes from IgA deficient patients to secrete IgA. J Clin Invest 1994; 94:97-104.

42. Zielen S, Bauscher P, Hofmann D et al. Interleukin-10 and immune restoration in common variable immunodeficiency. Lancet. 1993; 342:750-51.

43. Nonoyama S, Farrington M, Ishida H et al. Activated B cells from patients with common variable immunodeficiency proliferate and synthesize immunoglobulin. J Clin Invest 1993; 92:1282-87.

44. Farrington M, Grosmaire LS, Nonoyama S et al. CD40 ligand expression is defective in a subset of patients with common variable immunodeficiency. Proc Natl Acad Sci USA 1994; 91:1099-1103.

45. Pandrau D, Saeland S, Duvert V et al. Interleukin-4 inhibits in vitro proliferation of leukemic and normal human B cell precursors. J Clin Invest 1992; 90:1697-1706.

46. Banchereau J, de Paoli P, Vallé A et al. Long term human B cell lines dependent on interleukin 4 and anti-CD40. Science 1991; 251:70-72.

47. Chomarat P, Rissoan M-C, Banchereau J et al. Interferon γ inhibits Interleukin-10 production by monocytes. J Exp Med 1993; 177:523-27.

CHAPTER 5

REGULATION OF HUMAN MONOCYTE FUNCTIONS BY INTERLEUKIN-10

René de Waal Malefyt, Carl G. Figdor and Jan E. de Vries

INTRODUCTION

Monocytes and macrophages play an important role in the immune system through their cytotoxic activity, production of mediators and cytokines, ability to phagocytose native and opsonised antigens and presentation of antigenic peptides to T cells. These functions of monocytes/macrophages are influenced by the action of several cytokines: IFN-γ, GM-CSF, M-CSF and TNF-α activate monocytes, whereas IL-4, IL-13, TGFβ and IL-10 are potent monocyte deactivators. IL-10 plays a major role as dampener of immune responses through its deactivating potential of monocyte/macrophage functions.[1,2,3] The activities of IL-10 on monocyte functions are the subject of this chapter.

EFFECTS OF IL-10 ON MONOCYTE MORPHOLOGY

Monocytes isolated from peripheral blood adhere to plastic surfaces when placed in culture. Initially, this adherence is strong and it is accompanied by spreading of the cells (Fig. 5.1A). After 3 to 5 days of culture, some monocytes round up and detach. However, the majority of these cells cannot be removed by forceful pippetting or EDTA treatment. Culture of monocytes in the presence of IL-10 induces dramatic changes in their morphology and adhesion properties. Initially, monocytes will also adhere to plastic surfaces when cultured in the presence of IL-10, but most cells round up, detach and can be removed easily by pippetting after 3 to 5 days of culture (Fig. 5.1A). These cells are viable and will adhere to plastic surfaces when recultured in the absence of IL-10. In contrast, monocytes cultured in the presence of IL-4 or IL-13 will adhere strongly to plastic surfaces for periods longer than 5 days; it

Interleukin-10, edited by Jan E. deVries and René de Waal Malefyt.
© 1995 R.G. Landes Company.

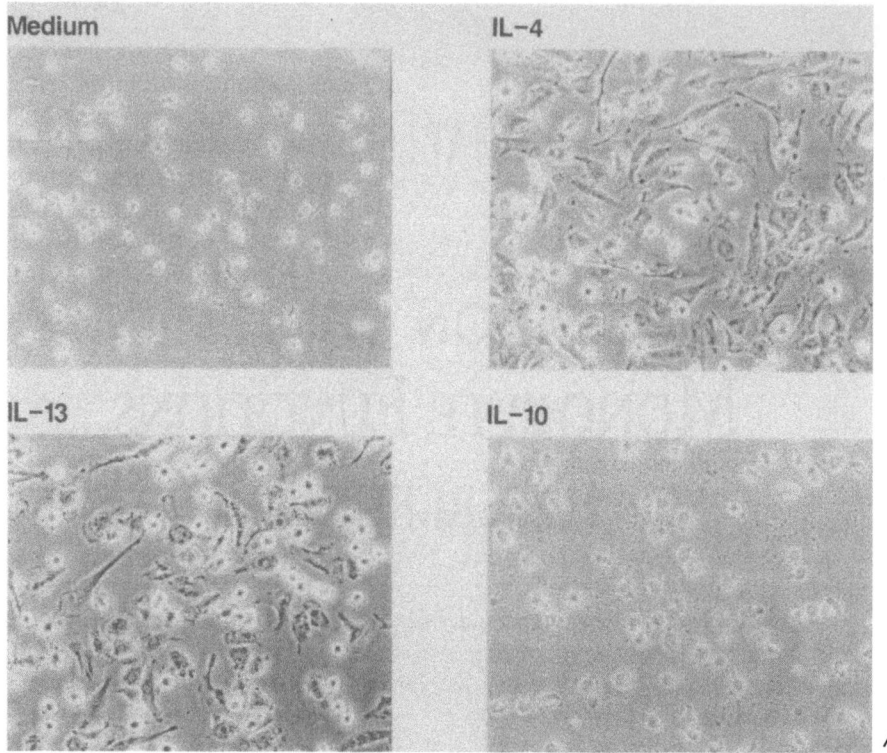

Figs. 5.1.A & B. IL-10 induces changes in the morphology of human monocytes. Human monocytes were isolated by centrifugal elutriation and cultured in the absence or presence of IL-4 (50ng/ml), IL-13 (50 ng/ml) or IL-10 (100 U/ml) for seven days and examined by A) light- and B) electronmicroscopy.

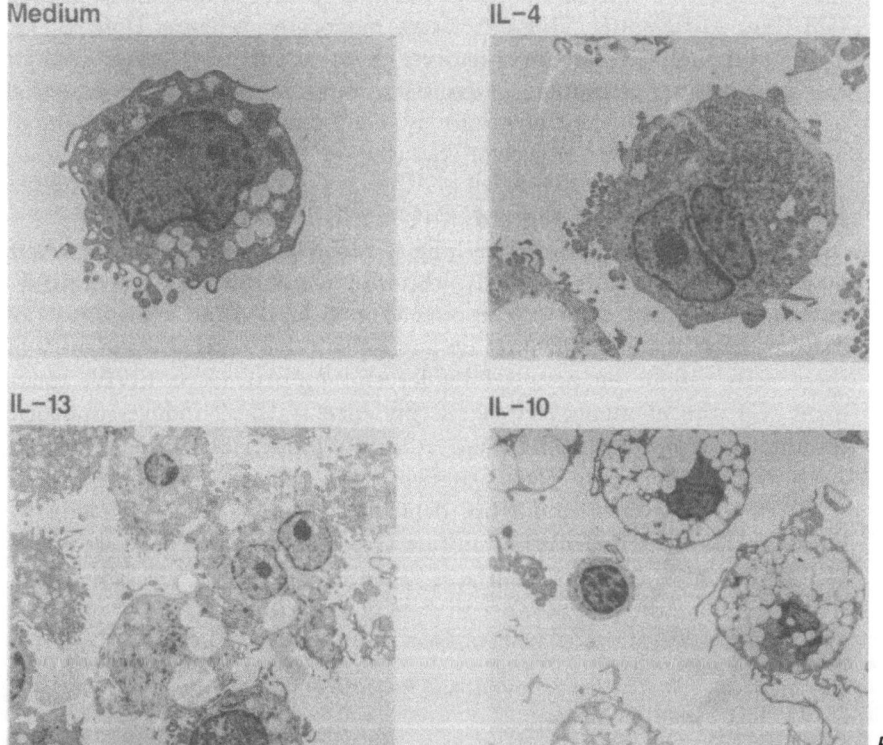

is not possible to remove these cells from the plastic by forceful pippetting or EDTA treatment. These cells form protrusions and long cytoplasmic processes which results in a dendritic appearance (Fig. 5.1A). Homotypic aggregations can also be observed in these cultures. Similar observations have been made with mouse macrophages cultured in the presence of IL-10, IL-4 or IL-13.[4]

To examine the morphology of monocytes cultured in the presence of IL-10, IL-4 or IL-13 in more detail, we analyzed their appearance by electron microscopy. Monocytes cultured in the presence of IL-10 for 5 days had a reduced cytoplasm which contained few secretory vacuoles and a poorly developed rough endoplasmic reticulum, indicating that these cells were not actively involved in protein synthesis (Fig. 5.1B). In addition, these cells contained large empty vacuoles in their cytoplasm. It is possible that these vacuoles contained water souble materials which were removed by the dehydration steps in the EM sample preparation process. In contrast, monocytes cultured in the presence of IL-4 or IL-13 had an enlarged cytoplasma containing numerous secretory vacuoles, and a well developed rough endoplasmic reticulum (Fig. 5.1B). It is clear that the dramatic effects of IL-10 on monocyte morphology will be indicative for the IL-10 induced changes in monocyte physiology and functions.

IL-10 MODULATES CLASS II MHC EXPRESSION ON MONOCYTES

IL-10 strongly downregulates the expression of MHC class II antigens on human monocytes. Freshly isolated human monocytes express relatively low levels of MHC class II antigens, which are spontaneously upregulated following culture overnight.[5] A significant induction of MHC class II antigens is observed following culture of the cells in the presence of IL-4, IL-13 or IFN-γ. IL-10 strongly downregulated the spontaneously and IL-4, or IL-13 or IFN-γ induced expression of MHC class II antigens.[6,7] Generally, the inhibi-

tory effects of IL-10 are more pronounced on the spontaneous and IFN-γ induced MHC class II expression as compared to that induced by IL-4 or IL-13. The expression of the three MHC class II subtypes, HLA-DR, HLA-DP and HLA-DQ antigens was inhibited to the same extend. Interestingly, the inhibition of MHC class II antigens was not at the level of steady state mRNA expression, since HLA-DRα–, HLA-DRβ–, and invariant chain mRNA expression was not affected by IL-10 (R. de Waal Malefyt, unpublished). This downregulation of MHC class II antigens on the cell surface of human monocytes by IL-10 may reduce the APC function of these cells and play an important role in the inhibitory effects of IL-10 on the antigen-specific activation of human T-cell clones.[6] In addition, enhanced levels of endogenous IL-10 may affect T-cell responses in autoimmune diseases, parasitic infections and graft versus host reactions (see below).

The effects of IL-10 on the constitutive or IFN-γ induced expression of MHC class II antigens on other cell types varied. IL-10 inhibited the upregulation of HLA-DR antigens on cultured human epidermal Langerhans cells,[8] and down-regulated MHC class II expression by corneal epithelial cells, keratocytes and endothelial cells as well as the inflammatory infiltrate in human herpes stromal keratitis.[9] In contrast, IL-10 did not affect MHC class II expression on EBV transformed B-cell lines,[6] or IFN-γ induced MHC class II expression on human astrocytes[10] and human dermal microvascular endothelial cells.[11] The effects of IL-10 on class II MHC expression on murine macrophages are controversial. It has been demonstrated that IL-10 did not affect IFN-γ induced MHC class II levels on a mouse macrophage cell line[4] or mouse macrophages.[12] However, others[10,13] demonstrated that IL-10 inhibited MHC class II expression on rat peritoneal macrophages and on mouse peritoneal macrophages and microglia cells following activation by IFN-γ. The reasons for these differences in the effects of IL-10 on MHC class II expression by these various cell types are not yet clear, but may be related to the absence or presence of IL-10 receptors on different cell types, the action of viral transactivators, the differentiation/activation state of the cells or specific culture conditions.

IL-10 MODULATES THE EXPRESSION OF ADHESION AND COSTIMULATORY MOLECULES ON MONOCYTES

Optimal T-cell activation and T-cell expansion requires triggering of the TCR and costimulatory signals provided by APC.[14] These costimulatory signals are mediated by interactions between CD28/CTLA-4, LFA-1 or CD2 molecules expressed on T cells, and B7/B70 (CD80/CD86), ICAM-1 or LFA-3 expressed on APC. IL-10 affects the expression of adhe-

sion and costimulatory molecules on monocytes. IL-10 inhibits the spontaneous upregulation of ICAM-1 (CD54) and B7 (CD80) antigens following culture in medium as well as the induction of CD80[15,16] (Fig. 5.2) by IFN-γ. CD80 is one of the ligands for CD28/CTLA-4 antigens on T cells, which are especially important for the activation of resting T cells.[17,18] A second ligand for CD28/CTLA-4, B70 (CD86), has recently been identified and cloned.[19,20] CD86 is constitutively expressed on freshly isolated human monocytes and dendritic cells and can be upregulated by IFN-γ. IL-10 inhibits both constitutive and IFN-γ induced expression of CD86 (Fig. 5.2). In murine models, IL-10 has also been shown to inhibit spontaneous and IFN-γ induced B7 expression by mouse macrophages.[12] Since CD86 is constitutively

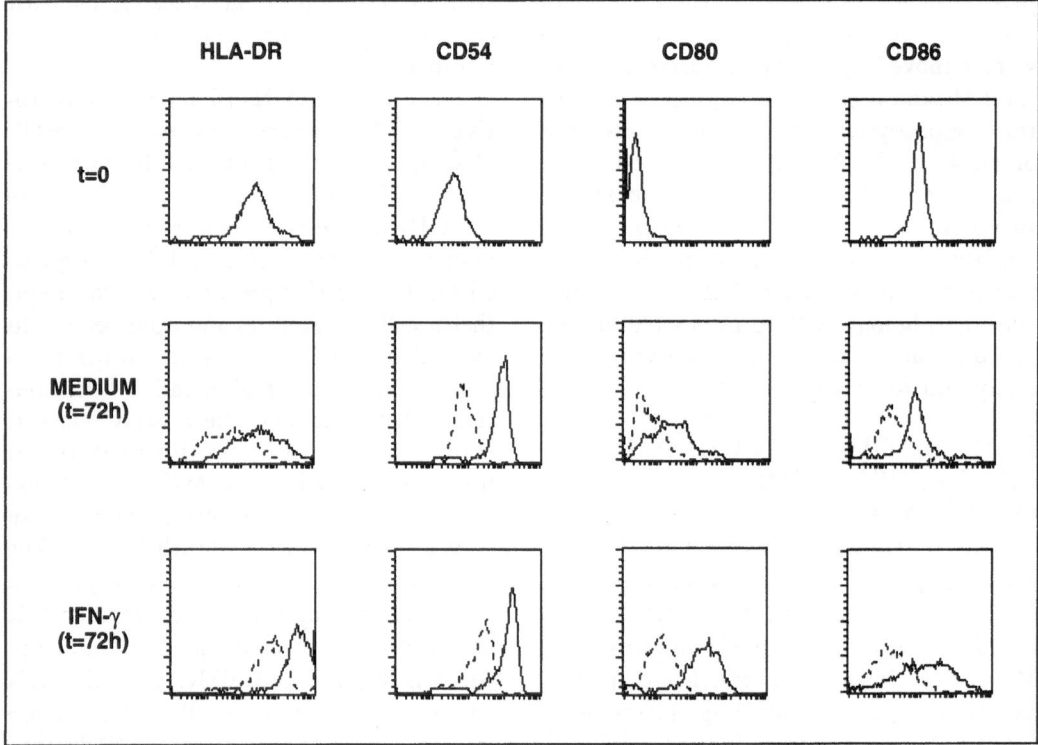

Fig. 5.2. Effects of IL-10 and IFN-γ on the expression of HLA-DR, CD54, CD80 and CD86 antigens by human monocytes. Human monocytes, isolated by negative selection using magnetic beads, were either immediately (t = 0), or following culture in medium or medium supplemented with IFN-γ (100 U/ml), in the absence or presence of IL-10 (100 U/ml) for 72 hrs, analyzed for expression of HLA-DR (L243), CD54 (ICAM-1) (LB-2), CD80 (B7) (L307) and CD86 (B70) (IT2) antigens. The x-axis represents fluorescence intensity (log scale) and the y-axis the relative cell number. Histograms of cells cultured in the presence of IL-10 are represented by the dotted lines.

expressed by monocytes and dendritic cells, it has been proposed that CD86/CD28 interactions are important for initiating activation of resting T cells leading to induction of CTLA-4 expression on the T cells and the production of cytokines, including IFN-γ, which in turn induces CD80 expression on the monocytes.[19] Interactions between CD86/CD80 on monocytes and CD28/CTLA-4 on T cells can subsequently result in enhanced T-cell activation and T-cell expansion, and amplification of the immune response. Inhibition of CD86 and CD80 expression on monocytes by IL-10 would strongly interfere with these processes, which would be consistent with its role as dampener of the immune response.[1]

IL-10 MODULATES MONOCYTE FC RECEPTOR EXPRESSION

IL-10 upregulates the expression of FcγRI (CD64)[21] to levels that are comparable to those induced by IFN-γ. Both IL-10 and IFN-γ induce tyrosine phosphorylation and formation of protein-DNA complexes between p91 (ISGF-3, interferon stimulated gene factor-3) and the GRR (IFN-γ response region) motif in the FcγRI promoter.[22] Binding of this IL-10 or IFN-γ activated complex to the GRR resulted in transcription. These data indicate that, at least for activation of the FcγRI promoter, IL-10 and IFN-γ signaling pathways are converging at the level of transcriptional activation. Interestingly, in mouse macrophages, IL-10 activated the binding of two latent receptor activated factors (RAFTs) to the GRR (gamma IFN activating site, GAS), which are related, but not identical to the p91 complex in human monocytes.[23] This may indicate that human monocytes and mouse macrophages differ in their potential to activate different transcription factors. IL-10 did not induce the expression of CD16 (FcγRIII) or CD32 (FcγRII), but it did prevent the downregulation of all three FcγRs which occurs when monocytes are cultured in the presence of IL-4 or IL-13.[7,21] Upregulation of CD64 on human monocyes by IL-10 correlated with an enhanced ADCC (anti-

body-dependent cytotoxicity) of these cells. However, IL-10 inhibited the cytotoxicity of monocytes and alveolar macrophages toward tumor cells, when monocytes were preactivated by LPS in the presence of IL-10.[24] The findings that the cytotoxicity of monocytes against tumor cells is inhibited by IL-10, whereas ADCC activity is enhanced, can be explained by fact that different cytotoxic mechanisms are resposible for these effector functions.[25]

IL-10 INHIBITS NO PRODUCTION BY MONOCYTES/MACROPHAGES

Nitric oxide plays an important role as mediator in a variety of biological processes, including vasodilatation, neurotransmission in the brain and peripheral nervous system, as well as macrophage cytotoxicity towards tumors, bacteria, fungi and parasites.[26] The production of NO by macrophages is not constitutive, but is induced following activation with MIF (migration inhibitory factor), IFN-γ or LPS. Pretreatment of mouse macrophages by IL-10 inhibited the IFN-γ and MIF dependent induction of nitric oxide synthase and the production of NO.[27-30] It has been shown that the production of NO by IFN-γ activated macrophages was dependent on endogenously produced TNF-α and that the inhibitory effect of IL-10 on IFN-γ induced NO production was mediated through the inhibition of this endogenously produced TNF-α.[31,32] Furthermore, IL-10 has been shown to synergize with IL-4 and TGF-β for inhibition of macrophage cytotoxic activity and NO production.[33] It is clear that inhibition of NO production significantly affects the cytotoxic potential of macrophages. Inhibition of NO production resulted in deminished cytotoxic activity of monocytes and enhanced the survival of several intracellular parasites, including *Toxoplasma gondii*,[29] *Leishmania donovani*,[30] *Trypanosoma cruzi*,[34] *Listeria mono-cytogenes*,[35] *Mycobacterium bovis*[32] and the fungus *Candida albicans*.[36] In addition, it has been described that enhanced NO levels following in vitro activation of macrophages by IFN-γ in the presence of anti-IL-10 mAbs protected susceptible mouse strains from death by *Candida albicans*.[37] Treatment of mice with neutralizing anti-IL-10 mAbs resulted in enhancement of early resistance against Listeria, but these mice failed to completely clear the infectious agents.[38] There are no reports on the effects of IL-10 on NO production by human monocytes In fact, it is possible that the gene encoding nitric oxide synthase is silent in human monocytes. However, the production of superoxide anion by human monocytes,[39] an other agent involved in the oxidative burst and monocyte cytotoxicity, was inhibited by viral IL-10. Furthermore, it has been demonstrated that cerebrospinal fluid of patients with bacterial meningitis contained IL-10[35] suggesting that IL-10 mediated inhibition of reactive oxidative intermediates (ROI) mediated killing mechanisms may play a role in the survival of bacteria in the cerebrospinal fluid in human.

LPS or IFN-γ induced NO production by a murine endothelial cell line was also inhibited by IL-10.[40]

IL-10 INHIBITS THE PRODUCTION OF CYTOKINES AND CHEMOKINES BY MONOCYTES AND NEUTROPHILS

Human monocytes are important producers of cytokines like IL-1α, IL-1β, IL-6, TNF-α and GM-CSF and chemokines such as IL-8 and MIP-1α. These cytokines play a crucial role in the recruitment and activation of granulocytes, monocytes, NK cells, T cells and B cells to sites of inflammation. IL-8 is a member of the C-X-C family of cytokines, which has chemotactic activities on neutrophils, whereas MIP-1α belongs to the C-C family of cytokines, which acts predominantly on macrophages and T cells.[41] IL-10 has potent inhibitory effects on the production of these pro-inflammatory cytokines.[42,43] IL-10 strongly inhibits the production of IL-1α, IL-1β, IL-6, TNF-α, IL-8 and MIP-1α by LPS, or LPS and IFN-γ activated human monocytes.[7,42] The inhibition of these cytokines was observed at the level of steady state

mRNA expression. Inhibition of IL-1β production by IL-10 has also been described in monocytes following activation by LPS or heat-killed *Staphylococcus aureus* at the level of IL-1β synthesis and not at the level of secretion of pro-IL-1β.[44] In addition to blocking these pro-inflammatory cytokines and chemo-kines, IL-10 inhibits the production of the hematopoietic growth factors GM-CSF, M-CSF and G-CSF by activated monocytes.[7,42,45] IL-10 inhibited the M-CSF production of monocytes activated with immobilized anti-CD45 mAbs and IL-1β at the transcriptional level[45] and the production of G-CSF protein and steady state mRNA levels by LPS activated monocytes.[42] It has to be noted that, although the primary functions of GM-, G- and M-CSF are promotion of growth of early hematopoietic progenitor cells, these growth factors, through their capacity to activate monocytes and thereby enhancing the production of pro-inflammatory cytokines and chemokines have a strong pro-inflammatory component. IL-10 also inhibited the production of IL-12. Both the synthesis of free IL-12 p40 chains and IL-12 p70 hetero-dimers by peripheral blood mononuclear cells activated by *Staphylococcus aureus* or LPS and the expression IL-12 p40 mRNA were inhibited.[46,47] In addition, IL-10 inhibited the expression of IL-12 p35 and IL-12 p40 mRNA by LPS activated monocytes.[7] Since IL-12 has been shown to be essential for directing Th1 responses[48,49] these results indicate that IL-10 by down-regulating IL-12 production may abrogate the development of Th1 and enhance Th2 pathways. Furthermore, IL-10 has been shown to inhibit the expression of IFN-α mRNA by LPS activated monocytes.[7] Administration of IFN-α induces down-regulation of Th2 responses that result in IgE synthesis in both mice[50] and human.[51] Therefore, it can be speculated that inhibition of IFN-α production by IL-10 again may favor the development of Th2 pathways.

IL-10 also inhibits the production of pro-inflammatory cytokines by human polymorphonuclear leukocytes (PMN). The production of IL-1α, IL-1β, IL-8 TNF-α,

MIP 1α and MIP 1β was inhibited by IL-10 following activation of PMN by LPS.[52-54] The IL-10 induced inhibition of IL-8 production by PMN is partly due to the inhibitory effects of IL-10 on IL-1 and TNF-α production,[52] which act as autrocrine stimulators of IL-8 production. Inhibition of these pro-inflammatory cytokines by IL-10 is at the level of steady state mRNA expression and evidence has been presented for effects of IL-10, both on mRNA stability and transcription.[53,54]

Similar inhibitory effects of IL-10 on the production of cytokines have been described on freshly isolated and periodate or thioglycolate induced peritoneal mouse macrophages.[4,27] IL-10 inhibited the production of IL-1α, IL-6 and TNF-α by LPS, or LPS and IFN-γ activated macrophage cell lines or macrophages at the level of steady state mRNA expression.[4] In addition, IL-10 has been shown to inhibit the production of TNF-α, IL-1α and IL-1β by LPS activated, periodate or thioglycolate elicited macrophages through degradation of cytokine mRNA.[27,55] This enhancement of cytokine mRNA degradation could be reversed by cyclohexamide, indicating that IL-10 induces a ribonuclease, or a protein which enhances the susceptibility of cytokine mRNA's to ribonucleases.[55] Thus, the inhibitory effects of IL-10 on cytokine mRNA levels in both, monocytes and PMN are mediated by effects on transcription and mRNA stability.[45,55,53] Also in the mouse IL-10 has been shown to inhibit the production of IL-12 by spleen or peritoneal cells of severe combined immunodeficiency (SCID) mice following activation with heat-killed *Listeria monocytogenes*[56] and by mouse peritoneal macrophages.[57]

Interestingly, IL-10 does not mediate a general downregulation of all cytokines by monocytes and PMN. The production of IL-1 receptor antagonist, a cytokine that binds type I and type II IL-1 receptors without activating them, is induced by IL-10 in monocytes,[58] whereas the production of IL-1ra in LPS activated monocytes[7,58] and neutrophils[59] is enhanced in the presence of IL-10. The upregulation of

IL-1ra in neutrophils by IL-10 was mediated through stabilization of IL-1ra mRNA.[59]

Apparently, not all cells that are able to produce pro-inflammatory cytokines are sensitive to the inhibitory effects of IL-10. The production of IL-6 and the induction of VCAM-1 expression by human umbilical vein endothelial cells (HUVEC) following activation by LPS or IL-1 was not inhibited by IL-10.[60,61] On the other hand, IL-10 enhanced the production of IL-6, and the chemokines MCP-1 and KC by a murine endothelioma cell line.[60] However, it needs to be established whether these HUVEC express functional receptors for IL-10. Collectively, these results indicate that IL-10, through its capacity to downregulate the production of pro-inflammatory cytokines and chemokines and its ability to enhance IL-1ra synthesis by monocytes and PMN, has not only potent anti-inflammatory activities, but may also prevent migration and accumulation of monocytes, PMN and T cells to the sites of inflammation. IL-10 shares these activities with IL-4 and IL-13.[42,62,63]

HUMAN MONOCYTES ARE ABLE TO PRODUCE IL-10 WHICH HAS AUTOREGULATORY ACTIVITIES

Human monocytes are able to produce large amounts of IL-10 following activation by LPS.[42] Interestingly, the production of IL-10 is late following activation relative to the rapid induction of the proinflammatory cytokines IL-1α, IL-1β, TNF-α and GM-CSF by LPS activated monocytes. IL-10 mRNA is detected at 8 hrs after activation and maximal levels are observed at 24 hrs. This late production of IL-10, which was also observed in activated T cells,[64] fits well with its biological role as dampener of immune responses.[1] The delayed production of IL-10 may be partly explained by the effects of TNF-α on IL-10 production. TNF-α has been shown to induce the expression of low levels of IL-10 by monocytes and, more importantly, anti-TNF-α antibodies could partly block the LPS induced expression of IL-10 by monocytes indicating that the production of TNF-α could be involved in a positive feedback regulation of IL-10 expression.[65] It is possible that IL-1β acts in a similar manner.[66] The production of IL-10 by monocytes is negatively affected by several other cytokines. First, IL-10 itself has an inhibitory effect on the expression of IL-10 mRNA expression in LPS activated human monocytes, indicating that it regulates its own expression.[42] The inhibitory effects of IL-10 on the expression of endogenously produced TNF-α could play a role in this autoregulatory effect of IL-10. Furthermore, IL-4, IL-13 and IFN-γ inhibit the production of IL-10 by LPS activated monocytes.[7,42,67] Inhibition of IL-10 mRNA expression by IFN-γ has also been described in the mouse.[4]

Endogenously produced IL-10 is biologically active and is able to inhibit the expression of cytokines by LPS activated monocytes. Endogenously produced IL-10 inhibited the production of IL-1α, IL-1β, IL-6, IL-8, IL-12, TNF-α, GM-CSF, G-CSF, MIP-1α and IFN-α. Activation of these cells in the presence of a neutralizing anti-IL-10 mAb, considerably enhanced the production of these cytokines in vitro. In addition, as expected, endogenously produced IL-10 was responsible for enhanced production of IL-1ra.[42] Endogenously produced IL-10 was also active as can auto-regulatory factor in the inhibition of MHC class II antigen expression on LPS activated monocytes.[42]

IL-10 INHIBITS ANTIGEN-PRESENTING FUCTIONS OF MONOCYTES

IL-10 strongly inhibits the antigen-specific proliferation of peripheral blood T cells and human T-cell clones belonging to Th0, Th1 and Th2 subsets towards soluble (protein) antigens, when monocytes are used as antigen-presenting cells.[6] In addition, IL-10 strongly inhibits the proliferation of alloreactive T-cell clones[68] and alloreactive T cells from peripheral blood in primary mixed leucocyte reactions when monocytes,[69] or Langerhans cells[8] are used as antigen-presenting cells. It is evident

that the effects of IL-10 on cytokine production and expression of MHC class II antigens, ICAM-1, CD80 and CD86 by monocytes will contribute to the mechanisms by which IL-10 inhibits T-cell activation. However, since T-cell clones reactive to soluble or alloantigens as well as resting T cells from peripheral blood have different activation requirements,[18,70,71] the importance of each of these mechanisms for the IL-10 mediated inhibition will depend on the type of antigen (allo-, soluble-, super,-), or mitogen used and on the differentiation state of the T cell (naive or activated). The inhibition of antigen-specific T-cell proliferation of human T-cell clones specific for soluble protein antigens (or peptides) seems to be primarily at the level of MHC class II/peptide association or expression on the monocyte since:

1. the inhibitory effects of IL-10 on T-cell proliferation are more pronounced at low concentrations of antigen, which are still able to optimally activate T-cell clones,[6]
2. the inhibitory effects of IL-10 could not be reversed by IL-1, IL-6, TNF-α or combinations of these cytokines,[6]
3. antigen specific proliferation of the T-cell clones could not be inhibited by CTLA-4 Ig fusion proteins, indicating that interactions between CD80/CD86 and CD28/CTLA-4 are not essential for the activation of T-cell clones[72] (R. de Waal Malefyt, unpublished)
4. IL-10 needs to be added at the beginning of the culture, indicating that it inhibits an early activation step[1] and
5. preincubation of monocytes with antigen and IL-10 resulted in an inability of T-cell clones to respond to MHC class II-peptide complexes as measured by increases in intracellular Ca^{2+} concentrations.[6]

Therefore, we proposed that IL-10 prevented the activation of T-cell clones by inhibiting MHC class II-peptide complex expression on cell surface of the monocytes. This hypothesis is confirmed by the observations that activation of T-cell clones by monocytes in the presence of IL-10 did not result in an upregulation of IL-2Rα chain expression (Fig. 5.3) or cytokine production by the T-cell clones[6,73,74] and that the inhibition of T-cell proliferation could not be reversed by adding physiological concentrations of IL-2 to these cultures.[6]

On the other hand, since blocking of CD80/CD86 and CD28/CTLA-4 interactions completely inhibit proliferative and cytotoxic responses in primary MLRs, it is likely that inhibition of CD80/CD86 on monocytes by IL-10 will be an important mechanism by which IL-10 inhibits the activation of resting T cells in allospecific reactions. However, the inhibitory effects of IL-10 on MHC class II expression may also play a role in the inhibition of these

alloresponses since alloantigen recognition is also dependent on antigen-processing and presentation by MHC class II molecules.[75,76] Furthermore, it has been shown that IL-10 inhibits the production of IL-1,[8] IL-6, TNF-α, GM-CSF and IFN-γ in primary MLR's when Langerhans cells,[8] or total PBMC were used as stimulators,[69] indicating that inhibition of these cytokines, which can act as costimulatory factors for T-cell proliferation, may also be involved in the inhibition of alloreactivity by IL-10. Clearly, the precise mechanisms by which IL-10 is able to inhibit the antigen presenting and accessory function of monocytes/macrophages is not completely resolved. It is possible that other, yet to be

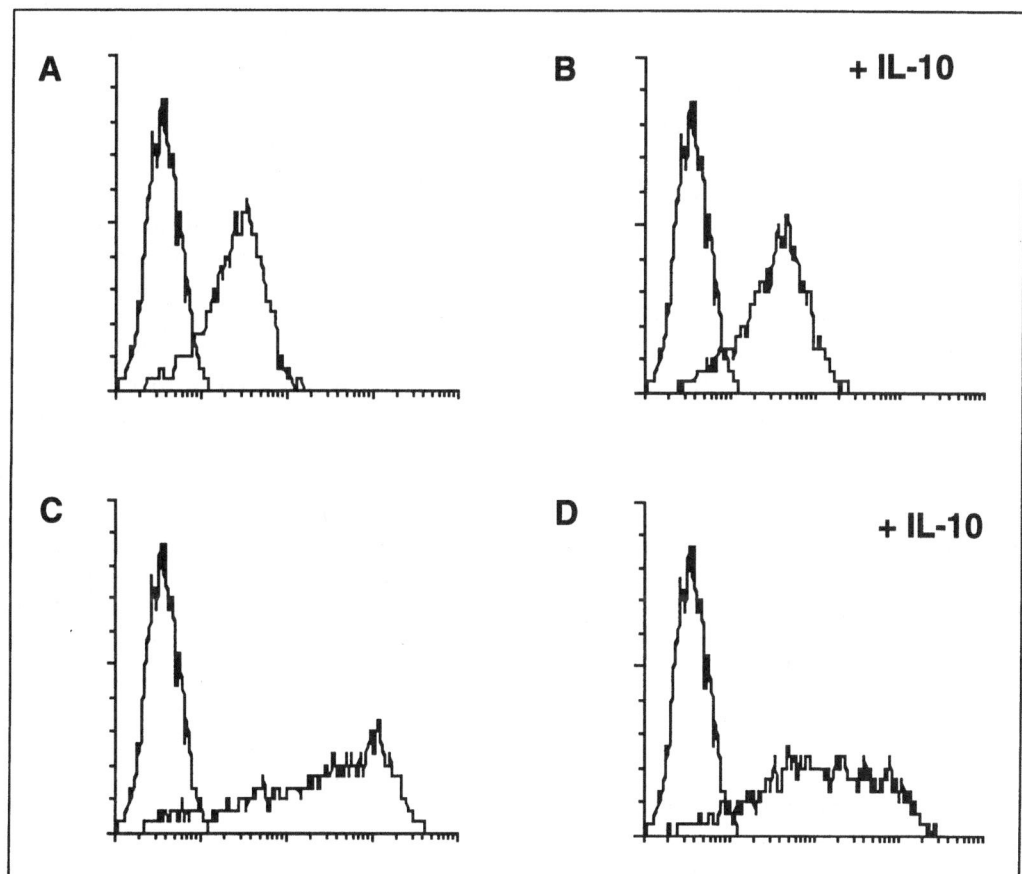

Fig. 5.3. Effects of IL-10 on the expression of CD25 by T-cell clone 827 following antigen-specific activation using HLA-matched monocytes as APC. Tetanus toxoid specific human T-cell clone 827 was cultured without (A, B) or with (C, D) HLA-matched human monocytes and tetanus toxoid in the absence or presence of IL-10 (100 U/ml) for 24 hrs and the expression of CD25 (IL-2Rα, TAC) was analyzed by immunofluorescence. The x-axis represents fluorescence intensity (log scale) and the y-axis the relative cell number. Histograms from cells stained with control antibody (nearest to the y-axis) are superimposed over the histograms of cells stained with anti-CD25 mAb.

defined, costimulatory cytokines or cell surface molecules also play a role in the mechanisms of IL-10 mediated inhibition of T-cell activation and proliferation. As discussed in chapter 4, IL-10 has direct effects on human T cells and specifically inhibits the production of IL-2.[71,77] Direct inhibition of IL-2 production by the responding T cells may also contribute to the inhibition of T-cell proliferation induced by antigen presented by monocytes. Inhibition of T-cell activation or proliferation by IL-10 has been demonstrated to occur in a number of pathological and physiological situations in vivo. IL-10 may account for the T-cell unresponsiveness observed in autoimmune diseases such as experimental autoimmune encephalomyelitis (EAE)[13,78,79] diabetes[80] and rheumatoid arthritis[66] and in parasitic infections such as visceral leishmaniasis[81-86] lepromatous lepracy,[87] schistosomiasis[88-90] and lymphatic filariasis.[91] Similarly IL-10 has been implicated in T-cell unresponsiveness in viral infections like HIV,[92,93] EBV[94,95] and murine AIDS.[96] IL-10 is also responsible for suppression of DTH responses following UV irradiation,[97,98] downregulation of contact hypersensitivity[99] and induction of hapten-specific tolerance.[100] In addition, IL-10 is associated with induction of tolerance and prevention of graft-versus-host disease following transplantation of hematopoietic stem cells,[68] heart allografts[101,102] and skin allografts.[103] These studies unequivocally show that IL-10 plays a role in dampening immune responses.[1]

IL-10 IS A POTENT ANTIINFLAMMATORY CYTOKINE

IL-10 may play an important role as dampener of inflammatory reactions through its effects on monocyte functions. As discussed here, IL-10 is able to strongly inhibit the production of pro-inflammatory cytokines and chemokines by monocytes following LPS activation.[4,42,43] Furthermore, IL-10 inhibits expression and induction of MHC class II antigens and other soluble, or membrane bound costimulatory molecules on APC, which results in abrogation of antigen-specific T-cell responses cytokine production by these cells.[6,12,16,57] In addition, IL-10 induces, or enhances the production of IL-1ra[7,58] and enhances the release of soluble TNF receptor p55 (TNFR) and TNFRp75 chains[104] by activated monocytes, thereby interfering with the receptor binding and effector functions of IL-1 and TNF-α. Furthermore, IL-10 inhibits the expression of tissue factor, which is a potent activator of procoagulant activity and may play a role in triggering disseminated intravascular coagulation (DIC) leading to multi organ dysfunction in endotoxemia or septicemia.[105,106] Finally, the inhibitory effects of IL-10 on the production of NO by macrophages, which is a potent vaso-dilatator, may contribute to its protective effects in shock. These anti-inflammatory activities of IL-10 are substantiated by observations made in in vivo studies, which indicated that IL-10 pretreatment of mice prevented their death from endotoxemia.[107,108] This correlated with inhibition of TNF-α and IFN-γ production.[109] In humans, IL-10 has also be implicated in the control of sepsis since IL-10 production was detected in the serum of patients with gram-positive or gram-negative septicaemia and higher levels were observed in patients with septic shock.[110] The important anti-inflammatory effects of IL-10 are also observed in mice which developed inflammatory bowel disease following targeting of their IL-10 gene, which could be prevented or delayed by administration of IL-10 from birth.[111] Taken together, the results obtained thus far indicate that IL-10 is an important regulator of immune and inflammatory responses by its potential to act as a monocyte deactivating factor.

ACKNOWLEDGMENTS

The authors would like to thank Dr. U. van Gorp (VITO, Mol, Belgium) for excellent E.M. analysis, Dr. L. Lanier and Dr. M. Azuma for generously providing mAbs and Ms. Jo Ann Katheiser for excellent secretarial assistance.

DNAX Research Institute of Molecular and Cellular Biology is supported by

Schering-Plough Corporation.

REFERENCES

1. de Waal Malefyt R, Yssel H, Roncarolo MG et al. Interleukin-10. Curr Opin Immunol 1992; 4:314-20.
2. Howard M, O'Garra A. Biological properties of interleukin 10. Immunol Today 1992; 13:198-200.
3. Moore KW, O'Garra A, de Waal Malefyt R et al. Interleukin-10. Annu Rev Immunol 1993; 11:165-90.
4. Fiorentino DF, Zlotnik A, Mosmann TR et al. IL-10 inhibits cytokine production by activated macrophages. J Immunol 1991; 147:3815-22.
5. Smith BR, Ault KA. Increase of surface Ia-like antigen expression on human monocytes independent of antigenic stimuli. J. Immunol. 1981; 127:2020.
6. de Waal Malefyt R, Haanen J, Spits H et al. Interleukin 10 (IL-10) and viral IL-10 strongly reduce antigen-specific human T cell proliferation by diminishing the antigen-presenting capacity of monocytes via downregulation of class II major histocompatibility complex expression. J Exp Med 1991; 174:915-24.
7. de Waal Malefyt R, Figdor CG, Huijbens R et al. Effects of IL-13 on phenotype, cytokine production, and cytotoxic function of human monocytes. Comparison with IL-4 and modulation by IFN-gamma or IL-10. J Immunol 1993; 151:6370-81.
8. Peguet NJ, Moulon C, Caux C et al. Interleukin-10 inhibits the primary allogeneic T cell response to human epidermal Langerhans cells. Eur J Immunol 1994; 24:884-91.
9. Boorstein SM, Elner SG, Meyer RF et al. IL-10 inhibition of HLA-DR expression in human herpes stromal keratitis. 1994; in press.
10. Frei K, Lins H, Schwerdel C et al. Antigen presentation in the central nervous system. The inhibitory effect of IL-10 on MHC class II expression and production of cytokines depends on the inducing signals and the type of cell analyzed. J Immunol 1994; 152:2720-8.
11. Vora M, Yssel H, de Vries JE et al. Anti-gen presentation by human dermal microvascular endothelial cells. Immuno-regulatory effect of IFN-γ and IL-10. J Immunol 1994; 152:5734-41.
12. Ding L, Linsley PS, Huang LY et al. IL-10 inhibits macrophage costimulatory activity by selectively inhibiting the up-regulation of B7 expression. J Immunol 1993; 151:1224-34.
13. Rott O, Fleischer B and Cash E. Interleukin-10 prevents experimental allergic encephalomyelitis in rats. Eur J Immunol 1994; 24:1434-40.
14. June CH, Bluestone JA, Nadler LM et al. The B7 and CD28 receptor families. Immunol Today 1994; 15:321- 31.
15. Willems F, Marchant A, Delville JP et al. Interleukin-10 inhibits B7 and intercellular adhesion molecule-1 expression on human monocytes. Eur J Immunol 1994; 24:1007-09.
16. Kubin M, Kamoun M, Trinchieri G. Interleukin-12 synergizes with B7/CD28 interaction in inducing efficient proliferation and cytokine production of human T cells. J Exp Med 1994; 180:211-22.
17. Linsley PS, Ledbetter JS. The role of the CD28 receptor during T cell responses to antigen. Annu Rev Immunol 1993; 11: 191-212.
18. Azuma M, Cayabyab M, Buck D et al. CD28 interaction with B7 co-stimulates primary allogeneic proliferative responses and cytotoxicity mediated by small, resting T lymphocytes. J Exp Med 1992; 175:353 -61.
19. Azuma M, Ito D, Yagita H et al. B70 antigen is a second ligand for CTLA-4 and CD28. Nature 1993; 366:76-78.
20. Freeman GJ, Gribben JG, Boussiotis VA et al. Cloning of B7-2: A CTLA-4 counter-receptor that costimulates human T cell proliferation. Science 1993; 262:909-11.
21. te Velde AA, de Waal Malefijt R, Huijbens RJ et al. IL-10 stimulates monocyte Fc gamma R surface expression and cytotoxic activity. Distinct regulation of antibody-dependent cellular cytotoxicity by IFN-gamma, IL-4, and IL-10. J Immunol 1992; 149:4048-52.
22. Larner AC, David M, Feldman GM et al. Tyrosine phosphorylation of DNA binding

proteins by multiple cytokines Science 1993; 261:1730-33.

23. Lehmann J, Seegert D, Strehlow I et al. IL-10 induced factors belonging to the p91 family of proteins bind to IFN-γ-responsive promoter elements. J Immunol 1994; 153:165-72.

24. Nabioullin R, Sone S, Mizuno K et al. Interleukin-10 is a potent inhibitor of tumor cytotoxicity by human monocytes and alveolar macrophages. J Leukoc Biol 1994; 55:437-42.

25. van Schie RCAA, Verstraten HGG, van de Winkel JGJ et al. Effect of recombinant IFN-γ (rIFN-γ) on the mechanism of human macrophage IgI FcRI-mediated cytotoxicity. J Immunol 1992; 148:169-76.

26. Lowenstein CJ, Snyder SH. Nitric Oxide, a novel biologic messenger. Cell 1992; 70:705-07.

27. Bogdan C, Vodovotz Y, Nathan C. Macrophage deactivation by interleukin 10. J Exp Med 1991; 174:1549-55.

28. Cunha FQ, Moncada S, Liew FY. Interleukin-10 (IL-10) inhibits the induction of nitric oxide synthase by interferon-gamma in murine macrophages. Biochem Biophys Res Commun 1992; 182:1155-59.

29. Gazzinelli RT, Oswald IP, James SL et al. IL-10 inhibits parasite killing and nitrogen oxide production by IFN-gamma-activated macrophages. J Immunol 1992; 148: 1792-96.

30. Wu J, Cunha FQ, Liew FY et al. IL-10 inhibits the synthesis of migration inhibitory factor and migration inhibitory factor-mediated macrophage activation. J Immunol 1993; 151:4325-32

31. Oswald IP, Wynn TA, Sher A et al. Interleukin 10 inhibits macrophage microbicidal activity by blocking the endogenous production of tumor necrosis factor alpha required as a costimulatory factor for interferon gamma-induced activation. Proc Natl Acad Sci USA 1992; 89:8676-80.

32. Flesch IEA, Hess JH, Oswald IP et al. Growth inhibition of mycobacterium bovis by IFN-γ stimulated macrophages: regulation by endogenous tumor necrosis factor-α and by IL-10. Int Immunol 1994; 6: 693-700.

33. Oswald IP, Gazzinelli RT, Sher A et al. IL-10 synergizes with IL-4 and transforming growth factor-beta to inhibit macrophage cytotoxic activity. J Immunol 1992; 148:3578-82.

34. Gazzinelli RT, Oswald IP, Hieny S et al. The microbicidal activity of interferon-gamma-treated macrophages against Trypanosoma cruzi involves an L-arginine-dependent, nitrogen oxide-mediated mechanism inhibitable by interleukin-10 and transforming growth factor-beta. Eur J Immunol 1992; 22:2501-06.

35. Frei K, Nadal D, Pfister HW et al. Listeria meningitis: identification of a cerebrospinal fluid inhibitor of macrophage listericidal function as interleukin 10. J Exp Med 1993; 178:1255-61.

36. Cenci E, Romani L, Mencacci A et al. Interleukin-4 and interleukin-10 inhibit nitric oxide-dependent macrophage killing of Candida albicans. Eur J Immunol 1993; 23:1034-38.

37. Romani L, Puccetti P, Mencacci A et al. Neutralization of IL-10 up-regulates nitric oxide production and protects susceptible mice from challenge with Candida albicans. J Immunol 1994; 152:3514-21.

38. Wagner D, Maroushek NM, Brown JF et al. Treatment with anti-interleukin-10 monoclonal antibody enhances early resistance to but impairs complete clearance of Listeria monocytogenes infection in mice. Infect and Immun. 1994; 62:2345-55.

39. Niiro H, Otsuka T, Abe M et al. Epstein-Barr virus BCRF1 gene product (viral interleukin 10) inhibits superoxide anion production by human monocytes. Lymphokine Cytokine Res 1992; 11:209-14.

40. Schneemann M, Schoedon G, Frei K et al. Immunovascular communication: activation and deactivation of murine endothelial cell nitric oxide synthase by cytokines. Immunol Lett 1993; 35:159-62.

41. Baggiolini M, Dewald B, Moser B. Interleukin-8 and related chemotactic cytokines CXC and CC chemokines. Adv Immunol 1994; 55:97-179.

42. de Waal Malefyt R, Abrams J, Bennett B et al. Interleukin 10 (IL-10) inhibits cytokine synthesis by human monocytes: an

autoregulatory role of IL-10 produced by monocytes. J Exp Med 1991; 174:1209-20.

43. Ralph P, Nakoinz I, Sampson JA et al. IL-10, T lymphocyte inhibitor of human blood cell production of IL-1 and tumor necrosis factor. J Immunol 1992; 148:808-14.

44. Chin J, Kostura MJ. Dissociation of IL-1 beta synthesis and secretion in human blood monocytes stimulated with bacterial cell wall products. J Immunol 1993; 151:5574-85.

45. Gruber MF, Williams CC, Gerrard TL. Macrophage-colony-stimulating factor expression by anti-CD45 stimulated human monocytes is transcriptionally up-regulated by IL-1 beta and inhibited by IL-4 and IL-10. J Immunol 1994; 152:1354-61.

46. D'Andrea A, Aste AM, Valiante NM et al. Interleukin 10 (IL-10) inhibits human lymphocyte interferon gamma-production by suppressing natural killer cell stimulatory factor/IL-12 synthesis in accessory cells. J Exp Med 1993; 178:1041-48.

47. Kubin M, Chow JM, Trinchieri G. Differential regulation of interleukin-12 (IL-12), tumor necrosis factor alpha, and IL-1 beta production in human myeloid leukemia cell lines and peripheral blood mononuclear cells. Blood 1994; 83:1847-55.

48. Hsieh CS, Macatonia SE, Tripp CS et al. Development of TH1 CD4+ T cells through IL-12 produced by Listeria-induced macrophages. Science 1993; 260:547-49.

49. Manetti R, Parronchi P, Guidizi MG et al. Natural kmiller cell stimulatory factor (interleukin-12 [IL-12]) induces T helper type 1 (Th1)-specific immune responses and inhibits the development of IL-4 -producing cells. J Exp Med 1993; 177:1199-1204.

50. Finkelman FD, Urban JF, Paul WF et al. Cytokine regulation of murine immunoglobulin isotype expression in vivo. In: IL-4: Structure and function, Spits H, ed. 1992:221-36.

51. Souillet G, Rousset F and de Vries JE. Alpha-interferon treatment of patient with hyper IgE syndrome. Lancet 1989; 1:1384.

52. Cassatella MA, Meda L, Bonora S et al. Interleukin 10 (IL-10) inhibits the release of proinflammatory cytokines from human polymorphonuclear leukocytes. Evidence for an autocrine role of tumor necrosis factor and IL-1 beta in mediating the production of IL-8 triggered by lipopolysaccharide. J Exp Med 1993; 178:2207-11.

53. Kasama T, Strieter RM, Lukacs NW et al. Regulation of neutrophil-derived chemokine expression by IL-10. J Immunol 1994; 152:3559-69.

54. Wang P, Wu P, Anthes JC et al. Interleukin-10 inhibits interleukin-8 production in human neutrophils. Blood 1994; 83:2678-83.

55. Bogdan C, Paik J, Vodovotz Y et al. Contrasting mechanisms for suppression of macrophage cytokine release by transforming growth factor-beta and interleukin-10. J Biol Chem 1992; 267:23301-08.

56. Tripp CS, Wolf SF, Unanue ER. Interleukin 12 and tumor necrosis factor alpha are costimulators of interferon gamma production by natural killer cells in severe combined immunodeficiency mice with listeriosis, and interleukin 10 is a physiologic antagonist. Proc Natl Acad Sci USA 1993; 90:3725-29.

57. Murphy EE, Terres G, Macatonia SE et al. B7 and interleukin-12 cooperate for proliferation and IFN-γ production by mouse Th1 clones that are unresponsive to B7 costimulation. J Exp Med 1994; 180:223-31.

58. Jenkins JK, Malyak M, Arend WP. The effects of interleukin-10 on interleukin-1 receptor antagonist and interleukin-1-beta production in human monocytes and neutrophils. Lymphokine and Cytokine Research 1994; 13:84-9.

59. Cassatella MA, Meda L, Gasperini S et al. Interleukin 10 (IL-10) upregulates IL-1 receptor antagonist production from lipopolysaccharide-stimulated human polymorphonuclear leukocytes by delaying mRNA degradation. J Exp Med 1994; 179:1695-99.

60. Sironi M, Munoz C, Pollicino T et al. Divergent effects of interleukin-10 on cytokine production by mononuclear phagocytes and endothelial cells. Eur J Immunol 1993; 23:2692-95.

61. Pugin J, Ulevitch RJ, Tobias PS. A critical role for monocytes and CD14 in endotoxin-induced endothelial cell activation. J Exp

Med 1993; 178:2193-200.

62. te Velde AA, Huijbens RJ, Heije K et al. IL-4 inhibits secretion of IL-1 beta, tumor necrosis factor alpha and IL-6 by human monocytes. Blood 1990; 76:1392-1399.

63. de Waal Malefyt R, Figdor CG, Huijbens R et al. Effects of IL-13 on phenotype, cytokine production, and cytotoxic function of human monocytes. J Immunol 1993; 151:6370-81.

64. Yssel H, de Waal Malefyt R, Roncarolo MG et al. IL-10 is produced by subsets of human CD4+ T cell clones and peripheral blood T cells. J Immunol 1992; 149: 2378-84.

65. Wanidworanun C and Strober W. Predominant role of tumor necrosis factor-alpha in human monocyte IL-10 synthesis. J Immunol 1993; 151:6853-61.

66. Katsikis PD, Chu CQ, Brennan FM et al. Immunoregulatory role of interleukin 10 in rheumatoid arthritis. J Exp Med 1994; 179:1517-27.

67. Chomarat P, Rissoan MC, Banchereau J et al. Interferon gamma inhibits interleukin 10 production by monocytes. J Exp Med 1993; 177:523-27.

68. Bacchetta R, Bigler M, Touraine JL et al. High levels of interleukin 10 production in vivo are associated with tolerance in SCID patients transplanted with HLA mismatched hematopoietic stem cells. J Exp Med 1994; 179:493-502.

69. Bejarano MT, de Waal Malefyt R, Abrams JS et al. Interleukin 10 inhibits allogeneic proliferative and cytotoxic T cell responses generated in primary mixed lymphocyte cultures. Int Immunol 1992; 4:1389-97.

70. Damle NK, Klussman K, Linsley PS et al. Differential costimulatory effects of adhesion molecules B7, ICAM-1, LFA-3 and V-CAM-1 on resting and antigen-primed CD4+ T lymphocytes. J Immunol 1992; 148:1985-94.

71. de Waal Malefyt R, Yssel H de Vries JE. Direct effects of IL-10 on subsets of human CD4+ T cell clones and resting T cells. Specific inhibition of IL-2 production and proliferation. J Immunol 1993; 150: 4754-65.

72. de Waal Malefyt R, Verma S, Bejarano M-

T et al. CD2/LFA3 or LFA-1/ICAM-1 but not CD28/B7 interactions can augment cytotoxicity by virus-specific CD8+ cytotoxic T lymphocytes. Eur J Immunol 1993; 23:418-24.

73. Del Prete G, De Carli M, Almerigogna F et al. Human IL-10 is produced by both type 1 helper (Th1) and type 2 helper (Th2) T cell clones and inhibits their antigen-specific proliferation and cytokine production. J Immunol 1993; 150:353-60.

74. Schlaak JF, Hermann E, Gallati H et al. Differential effects of IL-10 on proliferation and cytokine production of human gamma/delta and alpha/beta T cells. Scand J Immunol 1994; 39:209-15.

75. Rotzchke O, Falk K, Faath S et al. On the nature of peptides involved in T cell alloreactivity. J Exp Med 1991; 174: 1059-71.

76. Panina-Bordignon P, Gorradin G, Roosnek E et al. Recognition by class II alloreactive T cells of processed determinants from human serum proteins. Science 1991; 252:1548

77. Taga K, Mostowski H, Tosato G. Human interleukin-10 can directly inhibit T-cell growth. Blood 1993; 81:2964-71.

78. Kennedy MK, Torrance DS, Picha KS et al. Analysis of cytokine mRNA expression in the central nervous system of mice with experimental autoimmune encephalomyelitis reveals that IL-10 mRNA expression correlates with recovery. J Immunol 1992; 149:2496-505.

79. van der Veen RC and Stohlman SA. Encephalitogenic Th1 cells are inhibited by Th2 cells with related peptide specificity: relative roles of interleukin (IL)-4 and IL-10. J Neuroimmunol 1993; 48:213-20.

80. Pennline KJ, Roque-Gaffney E and Monahan M. Recombinant human IL-10 prevents the onset of diabetes in the nonobese diabetic mouse. Clin Immunol Immunopathol. 1994; 71:169-75.

81. Ghalib HW, Piuvezam MR, Skeiky YA et al. Interleukin 10 production correlates with pathology in human Leishmania donovani infections. J Clin Invest 1993; 92:324-29.

82. Holaday BJ, Pompeu MM, Jeronimo S et al. Potential role for interleukin-10 in the

immunosuppression associated with kala azar. J Clin Invest 1993; 92:2626-32.

83. Karp CL, el SS, Wynn TA et al. In vivo cytokine profiles in patients with kala-azar. Marked elevation of both interleukin-10 and interferon-gamma. J Clin Invest 1993; 91:1644-48.

84. Melby PC, Andrade NF, Darnell BJ et al. Increased expression of proinflammatory cytokines in chronic lesions of human cutaneous leishmaniasis. Infect Immun 1994; 62:837-42.

85. Powrie F, Menon S, Coffman RL. Interleukin-4 and interleukin-10 synergize to inhibit cell-mediated immunity in vivo. Eur J Immunol 1993; 23:3043-49.

86. Carvalho EM, Bacellar O, Brownell C et al. Restoration of IFN-γ production and lymphocyte proliferation in visceral leishmaniasis. J Immunol 1994; 152: 5949-56.

87. Sieling PA, Abrams JS, Yamamura M et al. Immunosuppressive roles for IL-10 and IL-4 in human infection. In vitro modulation of T cell responses in leprosy. J Immunol 1993; 150:5501-10.

88. Sher A, Fiorentino D, Caspar P et al. Production of IL-10 by CD4⁺ T lymphocytes correlates with down-regulation of Th1 cytokine synthesis in helminth infection. J Immunol 1991; 147:2713-16.

89. Flores VP, Chikunguwo SM, Harris TS et al. Role of IL-10 on antigen-presenting cell function for schistosomal egg-specific monoclonal T helper cell responses in vitro and in vivo. J Immunol 1993; 151:3192-98.

90. Villanueva P, Harris TS, Ricklan DE et al. Macrophages from schistosomal egg granulomas induce unresponsiveness in specific cloned th-1 lymphocytes in vitro and downregulate schistosomal granulomatous disease in vivo. J Immunol 1994; 152: 1847-55.

91. King CL, Mahanty S, Kumaraswami V et al. Cytokine control of parasite-specific anergy in human lymphatic filariasis. Preferential induction of a regulatory T helper type 2 lymphocyte subset. J Clin Invest 1993; 92:1667-73.

92. Clerici M, Wynn TA, Berzofsky JA et al. Role of interleukin-10 in T helper cell dys-

function in asymptomatic individuals infected with the human immunodeficiency virus. J Clin Invest 1994; 93:768-75.

93. Meyaard L, Schuitemaker H, Miedema F. T-cell dysfunction in HIV infection: anergy due to defective antigen-presenting cell function? Immunol Today 1993; 14:161-64.

94. Miyazaki I, Cheung RK, Dosch HM. Viral interleukin 10 is critical for the induction of B cell growth transformation by Epstein-Barr virus. J Exp Med 1993; 178:439-47.

95. Stewart JP, Rooney CM. The interleukin-10 homolog encoded by Epstein-Barr virus enhances the reactivation of virus-specific cytotoxic T cell and HLA-unrestricted killer cell responses. Virology 1992; 191:773-82.

96. Gazzinelli RT, Makino M, Chattopadhyay SK et al. CD4+ subset regulation in viral infection. Preferential activation of Th2 cells during progression of retrovirus-induced immunodeficiency in mice. J Immunol 1992; 148:182-88.

97. Rivas JM, Ullrich SE. Systemic suppression of delayed-type hypersensitivity by supernatants from UV-irradiated keratinocytes. An essential role for keratinocyte-derived IL-10. J Immunol 1992; 149:3865-71.

98. Ullrich SE. Mechanism involved in the systemic suppression of antigen-presenting cell function by UV irradiation. Keratinocyte-derived IL-10 modulates antigen-presenting cell function of splenic adherent cells. J Immunol 1994; 152:3410-16.

99. Ferguson TA, Dube P, Griffith TS. Regulation of contact hypersensitivity by interleukin 10. J Exp Med 1994; 179: 1597-604.

100. Enk AH, Saloga J, Becker D et al. Induction of hapten-specific tolerance by interleukin 10 in vivo. J Exp Med 1994; 179:1397-402.

101. Hancock W, Mottram PL, Purcell LJ et al. Prolonged survival of mouse cardiac allografts after CD4 or CD8 monoclonal antibody therapy is associated with selective intragraft cytokine protein expression: interleukin (IL)-4 and IL-10 but not IL-2 or interferon-gamma. Transplant Proc 1993; 25:2937-38.

102. Takeuchi T, Lowry RP, Konieczny B. Heart allografts in murine systems. The differen-

tial activation of Th2-like effector cells in peripheral tolerance. Transplantation 1992; 53:1281-94.

103. Gorczynski RM, Wojcik D. A role for non-specific (cyclosporin A) or specific (monoclonal antibodies to ICAM-1, LFA-1, and IL-10) immunomodulation in the prolongation of skin allografts after antigen-specific pretransplant immunization or transfusion. J Immunol 1994; 152:2011-19.

104. Leeuwenberg JFM, de Waal Malefyt R, Buurman WA. Slow release of soluble TNF receptors by monocytes in vitro. Leucocyte Typing 1994; V:in press

105. Pradier O, Gerard C, Delvaux A et al. Interleukin-10 inhibits the induction of monocyte procoagulant activity by bacterial lipopolysaccharide. Eur J Immunol 1993; 23:2700-03.

106. Ramani M, Ollivier V, Khechai F et al. Interleukin-10 inhibits endotoxin-induced tissue factor mRNA production by human monocytes. Febs Lett 1993; 334:114-6.

107. Howard M, Muchamuel T, Andrade S et al. Interleukin 10 protects mice from lethal endotoxemia. J Exp Med 1993; 177: 1205-08.

108. Gerard C, Bruyns C, Marchant A et al. Interleukin 10 reduces the release of tumor necrosis factor and prevents lethality in experimental endotoxemia. J Exp Med 1993; 177:547-50.

109. Marchant A, Bruyns C, Vandenabeele P et al. Interleukin-10 controls interferon-γ and tumor necrosis factor production during experimental endotoxemia. Eur J Immunol 1994; 24:1167-71.

110. Marchant A, Deviere J, Byl B et al. Interleukin-10 production during septicaemia. Lancet 1994; 343:707-08.

111. Kuhn R, Lohler J, Rennick D et al. Interleukin-10-deficient mice develop chronic enterocolitis. Cell 1993; 75:263-74.

CHAPTER 6

INTERLEUKIN-10: AN INHIBITOR OF MACROPHAGE-DEPENDENT STIMULATION OF TH1 CELLS

Anne O'Garra

INTRODUCTION

The original activity of interleukin-10 (IL-10), which led to its characterization at the molecular level was its ability to inhibit the induction of cytokine production by Th1 clones[1-3] and as a result of this it was given the name cytokine synthesis inhibitory factor (CSIF). IL-10, was first isolated from activated Th2-type, T helper cells, which produce cytokines promoting humoral and allergic type immune responses.[4] Its ability to inhibit the production of cytokines such as IFN-γ from Th1-type cells which mediate cell-mediated type immune responses, offered a possible explanation for why humoral and cell-mediated immune responses are often mutually exclusive.[5]

Upon isolation of a cDNA clone encoding the CSIF activity[2] and monoclonal antibodies directed against this cytokine,[6] it became evident that this molecule, now IL-10, was also responsible for two previously described activities, MCGF III[7,8] and B-TCGF.[9,10] These activities had been demonstrated in supernatants from a panel of mouse B cell lymphomas. They were found to costimulate the proliferation of mast cells and thymocytes, respectively.[7-10]

IL-10 INHIBITS MACROPHAGE BUT NOT B-CELL ANTIGEN PRESENTING FUNCTION FOR TH1 CLONES

Our finding that IL-10 was produced by B-cell lymphomas as well as by certain populations of activated, B-1 B cells[7,11] prompted us to

Interleukin-10, edited by Jan E. deVries and René de Waal Malefyt.

further investigate its mechanism of action to inhibit cytokine production by Th1 cells. The prediction was that B cell populations producing IL-10 would be unable to stimulate the production of cytokines by Th1 clones. However, this was not the case; B cells producing high levels of IL-10 induced significant cytokine production by Th1 clones.[12] Furthermore, addition of IL-10 or anti-IL-10 antibodies to these cultures of Th1 cells activated by antigen presented by B cell APC, had no effect on the cytokine levels produced.[12]

In contrast, when irradiated unseparated splenic or peritoneal cavity cells were used as APC for antigen- or mitogen-stimulated Th1 clones, IL-10 significantly inhibited the production of T cell-derived cytokines,[12] and anti-IL-10 mAbs often enhanced their production. IL-10 was not active when Th1 cells were stimulated with glutaraldhyde-fixed splenic APC, which indicated that its action involved regulation of APC function. IL-10 also inhibited cytokine synthesis by Th1 cells stimulated with irradiated splenic APC and the superantigen *Staphylococcus enterotoxin B*, which does not require processing. Purification of splenic or peritoneal cavity macrophages by FACS demonstrated that IL-10 mediated its effects by inhibition of macrophage APC stimulation of cytokine production by Th1 clones stimulated by antigen.[12] IL-10 also inhibited a IFN-γ-activated macrophage cell line from inducing cytokine production by Th1 clones.[12] Furthermore, stimulation of proliferation of human T helper cells by monocytes and antigen,[13] or of mouse T helper cells by macrophages and mitogen was also inhibited by IL-10.[14]

DIFFERENTIAL EFFECT OF IL-10 ON DENDRITIC CELL-INDUCED T CELL PROLIFERATION AND IFN-γ PRODUCTION

Since IL-10 selectively inhibited macrophage but not B cell APC function for Th1 cells it was of interest to determine its effects on the interdigitating dendritic cell (DC) APC function for induction of proliferation and IFN-γ production by both naive T cells and Th1 clones. DC APC have been established as potent inducers of naive T-cell proliferation both in vitro and in vivo.[15] IL-10 did not inhibit DC-induced proliferation of CD4 or CD8 T cells obtained from unimmunized mice,[14,16] or of Th1 cells,[16] although under the same conditions macrophage APC function for T-cell proliferation was inhibited. However, IL-10 inhibited DC induced IFN-γ production by both Th1 cells and CD4+ and CD8+ T cells from unprimed mice.[16]

Surprisingly, the DC were poor stimulators of Th1 cell proliferation, although as previously reported, potent stimulators of naive T-cell proliferation.[16] In contrast, maximal stimulation of Th1 clones was achieved using unseparated irradiated splenic APC which were at least two log orders of magnitude poorer at stimulating the proliferation of naive CD4+ and CD8+ T cells.[16] These phenomena can now be explained by the findings that antigen stimulated mouse Th1 clones[17] and mitogen activated human peripheral blood T cells[18] require signals delivered through B7-CD28 mediated interaction as well as IL-12 for the induction of maximal stimulation of IFN-γ production as well as significant induction of T-cell proliferation. This bears relevance to the mechanism of action of IL-10 (discussed below). This requirement of Th1 cells for B7 and IL-12 is in contrast to the antigen-specific stimulation of proliferation naive CD4+ T cells, which is maximal upon engagement of the TCR in the presence of costimulation via B7-CD28, and does not require the cooperation of IL-12.[17]

IL-10 INHIBITS THE DEVELOPMENT OF TH1 CELLS PRODUCING HIGH LEVELS OF IFN-γ

The transgenic mouse line DO10[19] expresses the αβ T-cell receptor (TCR) from the T-cell hybridoma DO11.10[20] reactive to the chicken ovalbumin peptide (OVA-323-339). The majority of the CD4+ T cells from this mouse line express the antigen-specific TCR clonotype from DO11.10 and can be identified with the anti-clonotypic monoclonal antibody (mAb) KJ1-26.[21] The majority of these CD4+ T cells have functional and phenotypic properties of naive T cells and thus can be used to study the development of T helper subsets. When irradiated BALB/c or SCID splenocytes were used as APC to prime these TCR-transgenic CD4+ T cells to the ovalbumin peptide antigen they developed into a Th0 phenotype producing low levels of IL-4 and IFN-γ upon subsequent restimulation with OVA plus splenic APC.[22] Neutralization of IL-10 during primary stimulation, by the addition of a monoclonal antibody directed against the cytokine, SXC-1,[6] resulted in development of a Th1 phenotype, with strong IFN-γ production and low to undetectable IL-4 and IL-5 production upon restimulation. This effect of anti-IL-10 antibodies was only observed with splenocyte APC and not with a B lymphoma, TA3, in keeping with the effects of IL-10 on the APC function for upregulation of IFN-γ by already differentiated Th1 clones (see above). In contrast to the APC-dependent effects obtained with anti-IL-10 antibodies, anti-IL-4 antibodies allowed the development of Th1 cells producing high levels of IFN-γ upon restimulation, independent of the APC used to prime the T cells. Furthermore, IL-4 was able to drive the development of Th2 cells producing high levels of IL-4 upon restimulation with antigen, whereas IL-10 did not and was only able to inhibit the development of a Th1 phenotype, in this situation when present endogenously in priming cultures.[22]

We have more recently found that the addition of heat-killed *Listeria monocytogenes* (*Listeria*) during primary activation of OVA-specific T cells resulted in enhanced IFN-γ production in primary cultures and rapid development of a Th1 phenotype.[23,24] The action of *Listeria* was via the macrophage to drive the development of Th1 cells producing significant levels of IFN-γ, provided the naive CD4+ T cells were appropriately primed to specific antigen (OVA) with dendritic cells.[23,24] The mechanism of *Listeria*-induced IFN-γ production

and Th1 development was dependent on IL-12. Moreover, IL-12, previously described as an NK-cell stimulating factor,[25] could completely replace the Th1 inducing capacity of *Listeria*-activated macrophages.[23] IL-12 is now recognized as a dominant Th1 inducing factor, both in mouse[23] and human[26] systems.

IL-10 inhibited both the macrophage-dependent *Listeria*-induced IFN-γ production from naive T cells and development of a Th1 phenotype.[23,24] However, IL-10 had no effect on IL-12-induced Th1 development and IFN-γ production[23] suggesting that its action was by blocking IL-12 production by macrophages. We and others have now confirmed that this action of IL-10 was by down-regulation of macrophage-derived IL-12[17,27] (see below).

MECHANISM OF ACTION OF IL-10 FOR INHIBITION OF MACROPHAGE APC FUNCTION

A possible mechanism of action of IL-10, to inhibit macrophage APC function for Th1 clones, was to interfere with TCR-class II MHC peptide interactions. IL-10 did not down-regulate the IFN-γ induced Class II levels on a mouse macrophage cell line whose APC function for Th1 cells was down-regulated by this factor,[12] in contrast to the data of others that IL-10 down-regulates Class II expression on human monocytes.[13] To support an alternative mechanism of action for IL-10 on murine macrophages, IL-10 also inhibited IL-2-induced IFN-γ production by Th1 clones, in an antigen-free system requiring only the presence of adherent accessory cells.[12] These data suggested that IL-10 inhibited macrophage accessory cell function which is independent of TCR-class II MHC-peptide interactions.

Further studies on the direct effect of IL-10 on mouse macrophages revealed that this cytokine was a potent inhibitor of LPS-induced monokine production,[28] and de Waal Malefyt et al, showed that this inhibitory effect of IL-10 was also evident on human monocytes.[29] Indeed IL-10 was shown to downregulate a number of macrophage functions, including the production of IL-1, IL-6, TNF-α, GM-CSF and G-CSF, as well as the generation of reactive nitrogen intermediates,[28-31] which are effector molecules involved in the elimination of intracellular and extracellular parasites. Although IL-10 downregulated the production of monokines including IL-1, IL-6 and TNF-α, none of these were able to overcome the effect of IL-10 to inhibit macrophage APC function for mouse Th1 clones.[28] However, it was still possible that IL-10 downregulated another costimulatory activity needed for optimal Th1 cytokine secretion in response to macrophages and antigen. However, using supernatants obtained from macrophage cell lines known to be inhibitable by IL-10, which had been activated in a variety of ways, it was not possible to overcome the inhibitory effect of IL-10 on macrophage and antigen-dependent stimulation of Th1 cells.[28] The alternative that IL-10 induced the production of a macrophage-derived inhibitor which acted directly on the Th1 cells to inhibit cytokine production, was ruled out using cell mixing experiments[28] and was also supported by similar data from Ding and Shevach.[14] These findings suggested that either: (i) IL-10 did not mediate its effects by downregulation of a soluble costimulator; (ii) that such a costimulator was highly labile, absorbed during isolation, or (iii) such a costimulator was present at very low concentrations and active as a short-range acting factor; (iv) or finally that IL-10 may inhibit the expression of a membrane-bound costimulator. In fact it is now clear that IL-10 can mediate its effects on the downregulation of macrophage APC or accessory function by both down-regulation of a soluble mediator, IL-12,[17,27] and a membrane-bound co-stimulator, B7.[18,32] We[17] and others studying the effects of IL-10 on activated monocytes in human systems,[27] have shown that IL-10 can inhibit the expression of IL-12 by macrophages activated with IFN-γ. Furthermore, in our hands IL-12 can overcome the effects of mouse IL-10 to inhibit macrophage-dependent stimulation of Th1 cells.[17]

In summary, we have demonstrated that IL-10 is a potent inhibitor of mouse macrophage function, and inhibits macrophage-dependent stimulation of IFN-γ from both naive CD4+ T cells and Th1 clones, as well as the development of a Th1 phenotype from naive CD4+ cells.[12,23,24,28] Furthermore, a major mechanism of IL-10 to inhibit macrophage-dependent stimulation of Th1 cell development,[23] as well as IFN-γ production from mouse and human CD4+ T cells[17,27] is by inhibition of IL-12 production.

References

1. Fiorentino DF, Bond MW, Mosmann TR. Two types of mouse helper T cell. IV. Th2 clones secrete a factor that inhibits cytokine production by Th1 clones. J Exp Med 1989; 170:2081-95.

2. Moore KW, Vieira P, Fiorentino DF et al. Homology of cytokine synthesis inhibitory factor (IL-10) to the Epstein Barr Virus gene BCRFI. Science 1990; 248:1230-34.

3. Vieira P, de Waal-Malefyt R, Dang M-N et al. Isolation and expression of human cytokine synthesis inhibitory factor (CSIF/IL10) cDNA clones: homology to Epstein-Barr virus open reading frame BCRFI. Proc Natl Acad Sci USA 1991; 88:1172-76.

4. Mosmann TR, Bond MW, Coffman RL et al. T-cell and mast cell lines respond to B-cell stimulatory factor 1. Proc Natl Acad Sci USA 1986; 83:5654-58.

5. Parish CR. The relationship between humoral and cell-mediated immunity. Transplant Rev 1972; 13:35-66.

6. Mosmann TR, Schumacher J, Fiorentino DF et al. Isolation of monoclonal antibodies specific for IL4, IL5, IL6, and a new Th2-specific cytokine (IL-10), cytokine synthesis inhibitory factor, by using a solid phase radioimmunoadsorbent assay. J Immunol 1990; 145:2938-45.

7. O'Garra A, Stapleton G, Dhar V et al. Production of cytokines by mouse B cells: B lymphomas and normal B cells produce Interleukin-10. Int Immunol 1990; 2: 821-32.

8. Thompson-Snipes L, Dhar V, Bond MW et

al. Interleukin-10: a novel stimulatory factor for mast cells and their progenitors. J Exp Med 1991; 173:507-10.

9. Suda T, O'Garra A, MacNeil I et al. Identification of a novel thymocyte growth promoting factor derived from B cell lymphomas. Cell Immunol 1990; 129: 228-40.

10. Willoughby PB, Jennette JC, Haughton G. Analysis of a murine B cell lymphoma, CH44, with an associated non-neoplastic T cell population. I. Proliferation of normal T lymphocytes is induced by a secreted product of the malignant B cells. Am J Pathol 1988; 133:507-15.

11. O'Garra A, Chang R, Go N et al. Ly1 B (B-1) cells are the main source of B-cell derived IL-10. Eur J Immunol 1992; 22: 711-17.

12. Fiorentino DF, Zlotnik A, Vieira P et al. IL-10 acts on the antigen-presenting cell to inhibit cytokine production by Th1 cells. J Immunol 1991; 146:3444-51.

13. de Waal Malefyt R, Haanen J, Yssel H et al. IL-10 and v-IL-10 strongly reduce antigen specific human T cell responses by diminishing the antigen presenting capacity of monocytes via down-regulation of class II MHC expression. J Exp Med 1991; 174:915-24.

14. Ding L, Shevach EM. IL-10 inhibits mitogen-induced T cell proliferation by selectively inhibiting macrophage costimulatory function. J Immunol 1992; 148:3133-39.

15. Steinman RM, Young JW. Signals arising from antigen-presenting cells. Curr Opin Immunol 1991; 3:361-72.

16. Macatonia SE, Doherty TM, Knight SC et al. Differential effect of Interleukin-10 on dendritic cell-induced T cell proliferation and Interferon-γ production. J Immunol 1993; 150:3755-65.

17. Murphy EE, Terres G, Macatonia SE et al. B7 and interleukin-12 cooperate for proliferation and IFN-γ production by mouse Th1 clones that are unresponsive to B7 costimulation. J Exp Med 1994; 180: 223-31.

18. Kubin M, Kamoun M, Trinchieri G. Interleukin-12 synergizes with B7/CD28 interaction in inducing efficient proliferation and cytokine production of human T cells. J

Exp Med 1994; 180:212-20.

19. Murphy KM, Heimberger AB, Loh DY. Induction by antigen of interthymic apoptosis of CD4+ CD8+ TCR-lo thymocytes in vivo. Science 1990; 250:1720-22.

20. Kappler JW, Skidmore B, White J et al. Antigen-indivisable, H-2 restricted, interleukin-2 producing T cell hybridomas lack of independent antigen and H-2 recognition. J Exp Med 1981; 153:1198-1214.

21. Marrack P, Shimonkevitz R, Hannum C et al. The Major Histocompatibility Complex-restricted antigen receptor on T Cells. J Exp Med 1983; 158:1635-46.

22. Hsieh C-S, Heimberger AB, Gold JS et al. Differential regulation of T helper phenotype development by IL-4 and IL-10 in an αβ-transgenic system. Proc Natl Acad Sci USA 1992; 89:6065-69.

23. Hsieh C-S, Macatonia SE, Tripp CS et al. Development of Th1 CD4+ T cells through IL-12 produced by Listeria-induced macrophages. Science 1993; 260:547-49.

24. Macatonia SE, Hsieh C-S, Murphy KM et al. Dendritic cells and macrophages are required for Th1 development of CD4+ T cells from αβ TCR transgenic mice: IL-12 substitution for macrophages to stimulate IFN-γ production is IFN-γ-dependent. Int Immunol 1993; 5:1119-28.

25. Kobayashi M, Fitz L, Ryan M et al. Identification and purification of natural killer cell stimulatory factor (NKSF), a cytokine with multiple biologic effects on human lymphocytes. J Exp Med 1989; 170:827-45.

26. Manetti R, Parronchi P, Guidizi MG et al. Natural killer cell stimulatory factor (interleukin 12 [IL-12]) induces T helper type 1 (Th1)-specific immune responses and inhibits the development of IL-4-producing cells. J Exp Med 1993; 177:1199-1204.

27. D'Andrea AD, Aste-Amezaga M, Valainte NM et al. Interleukin-10 inhibits lymphocyte IFN-γ production by suppressing natural killer cell stimulatory factor/interleukin-12 synthesis in accessory cells. J Exp Med 1993; 178:1041-48.

28. Fiorentino DF, Zlotnik A, Mosmann TR et al. IL-10 inhibits cytokine production by activated macrophages. J Immunol 1991; 147:3815-22.

29. de Waal Malefyt R, Abrams J, Bennett B et al. IL-10 inhibits cytokine synthesis by human monocytes: An autoregulatory role of IL-10 produced by monocytes. J Exp Med 1991; 174:1209-20.

30. Bogdan C, Vodovotz Y, Nathan C. Macrophage deactivation by interleukin-10. J Exp Med 1991; 174:1549-55.

31. Gazzinelli RT, Oswald IP, James SL et al. IL-10 inhibits parasite killing and nitric oxide production by IFN-γ activated macrophages. J Immunol 1992; 148:1792-96.

32. Ding L, Linsley PS, Huang LY et al. IL-12 inhibits macrophage costimulatory activity by selectively inhibiting the upregulation of B7 expression. J Immunol 1993; 151:1224-34.

CHAPTER 7

IL-10 AND BONE FORMATION/HEMATOPOIESIS

Peter Van Vlasselaer

INTRODUCTION

Considering the anatomical localization of bone and bone marrow it is not surprising that both tissues are composed of cells from hematopoietic as well as from stromal origin. Indeed, apart from the stroma-derived osteoblasts and osteocytes, which are characterized by their ability to produce a mineralized matrix[1,2] bone also contains osteoclasts. Osteoclasts belong to the hematopoietic lineage and have a lot in common with macrophages and characteristically resorb bone.[3,4] The hematopoietic compartment of the bone marrow on the other hand, is functionally and structurally supported by a microenvironment of stromal elements including adipocytes, fibroblasts, endothelial cells and undifferentiated mesenchymal cells. These mesenchymal cells represent a reservoir of "uncommitted", self-renewing cells which differentiate into various stromal elements, including bone.[2,5] Under normal conditions, hematopoiesis and stromal differentiation require complex sequences of cellular events that are modulated by site-specific and cell-specific signals capable of initializing and promoting the recruitment and proliferation of the appropriate cells at the right time. These signals are mediated by hormones, prostaglandins, growth factors and cytokines which either reside in the bone matrix or in the bone marrow. The goal of this chapter is to review the effects of interleukin-10 (IL-10) on bone formation and hematopoiesis. IL-10 was initially described as cytokine synthesis inhibiting factor (CSIF) based on its potential to block the synthesis of cytokines produced by type 2 helper T cells.[6]

IL-10 AND BONE FORMATION

THE ORIGIN AND CHARACTERISTICS OF OSTEOPROGENITOR CELLS

Bone marrow stroma forms a network of fibroblasts, adipocytes, endothelial cells and mesenchymal cells that supports and regulates hemato-

Interleukin-10, edited by Jan E. deVries and René de Waal Malefyt.
© 1995 R.G. Landes Company.

poiesis and harbors cells that give rise to the osteogenic lineage. The presence of osteoprogenitor cells in the marrow stroma is illustrated by the fact that bone marrow differentiates into bone ossicles when transplanted under the kidney or when cultured in intraperitoneally implanted diffusion chambers.[7,8] In addition, a number of immortalized and transfected cell lines were generated from bone marrow stroma, which elicit osteogenic characteristics when cultured in vitro or when transplanted in vivo.[9,10] Recent studies in the mouse showed that osteoprogenitor cells are low density cells (1.066-1.067 g/ml), which represent 0.00045% of the cells in normal bone marrow and 0.0057% in marrow treated with the chemotoxic drug 5-fluorouracil (5-FU). Moreover, these cells form bone nodules in culture and bind to both wheat germ and soybean agglutinin.[11] Since 5-FU treatment depletes more than 95% of the circulating cells in vivo without reducing the frequency of osteoprogenitor cells, it is fair to say that the latter cells reside in the bone marrow in a quiescent state. Immunohistochemistry and in situ hybridization showed that mesenchymal cells of the bone marrow are uncommitted to the bone lineage since they do not produce or express bone related proteins. However, when cultured in the presence of β-glycerophosphate and vitamin C, mesenchymal cells quickly differentiate into bone. During this process they sequentially secrete alkaline phosphatase (ALP), collagen type I and osteocalcin, and finally they form a mineralized matrix which contains the bone specific mineral, hydroxyapatite.[11] Immunophenotypical analysis in combination with cell sorting revealed that uncommitted murine mesenchymal cells display a high forward and perpendicular light scatter profile. In addition, these cells bind wheat germ agglutinin and the monoclonal antibodies Sca-1, KM16, Sab-1 and Sab-2. Whereas, Sca-1 is commonly expressed on hematopoietic stem cells, KM16, Sab-1 and Sab-2 are specific for the stromal lineage.[12,13] On the other hand, osteoprogenitor cells do not express the hematopoietic lineage markers Gr-1, B220, L3T4, Lyt2, Thy-1, Mac-1, Mac-2 and Mac-3.[43] A comparable observation was made in the human bone marrow where osteoprogenitor cells express the hematopoietic and stromal marker CD34 and STRO-1[14] respectively, but none of the lineage markers (unpublished observation). Hence, it appears that mesenchymal cells express markers that are co-expressed by cells of the hematopoietic lineage. This indicates that hematopoietic and stromal cells may originate from a common precursor. Recent data revealed that at least the fetal human bone marrow contains a CD34+ CD38-DR- cell type, which gives progeny to both hematopoietic and stromal cells.[16] However, personal observations showed that Sca-1+WGA+ and CD34+ CD38- cells from mature murine and human bone marrow are unable to differentiate into either hematopoietic or stromal cells respectively (unpublished observation).

In conclusion, a vast amount of data indicate that osteoprogenitor cells originate from uncommitted, pluripotent mesenchymal cells which represent a minor population in the bone marrow.

IL-10 REGULATES BONE PROTEIN SYNTHESIS AND EXTRACELLULAR MATRIX FORMATION

In vitro and in vivo bone formation is characterized by an increased proliferation of osteogenic cells, followed by subsequent synthesis of ALP, collagen type I and osteocalcin which finally leads to the formation of a mineralized matrix. We developed an in vitro model for bone formation to screen the effect of different cytokines on the sequential events, which govern osteogenic differentiation.[11] When IL-10 was administered to this culture system, it significantly decreased the synthesis of ALP, collagen type I and osteocalcin (Fig. 7.1).[16] Surprisingly, this effect was not preceded by an overall suppression of DNA synthesis or by a reduction of the growth, or the colony size, of the fibroblast colony forming cells (CFU-F), which are believed

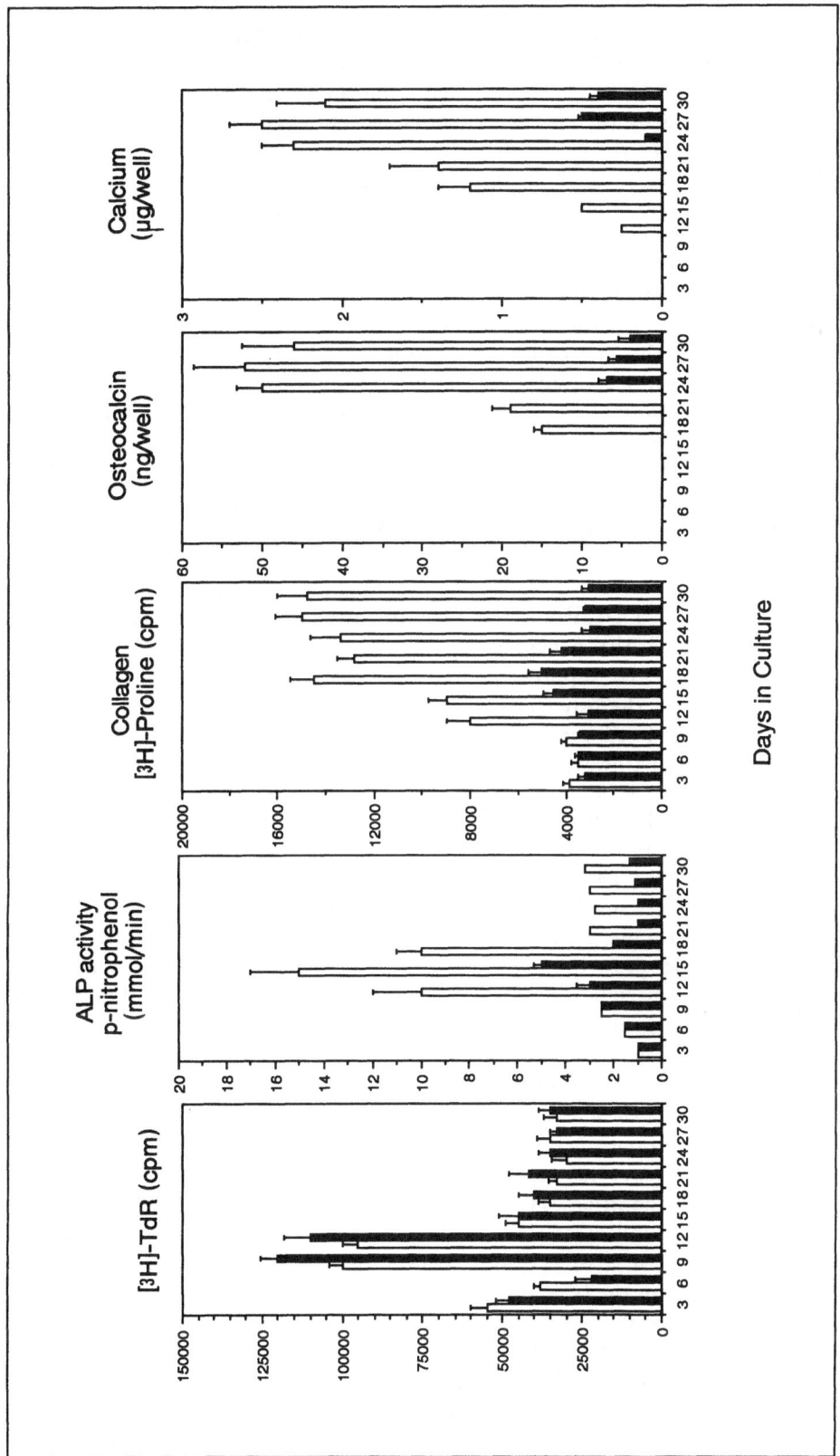

Fig. 7.1. Effect of IL-10 on the temporal [³H]-thymidine incorporation, ALP, collagen and osteocalcin synthesis and mineralization in cultures of 5-FU treated mouse bone marrow. White bars and black bars represent untreated and IL-10 treated cultures, respectively. The results represent the mean + SD of quadruplicate cultures.

to harbor the osteoprogenitor cells. Since the suppressive effect of IL-10 coincided with a reduced expression of bone protein mRNA, this suggests that IL-10 merely suppresses bone protein synthesis. However, this view is not supported by the observation that IL-10 suppressive activity only occurred when it was added to the culture system before the earliest sign of osteogenic commitment, e.g. ALP synthesis. When added beyond that point in time, IL-10 was not able to reduce bone protein synthesis or mineralization. Furthermore, IL-10 failed to suppress the constitutive bone protein synthesis by the osteoblastic cell lines MC3T3-E1[17] and MN7.[10] In other words, mature bone cells do not respond to IL-10. In conclusion, it appears that IL-10 affects the early events, which monitor the onset of osteogenic differentiation of mesenchymal cells.[16]

THE "IL-10—TGF-β 1 LOOP"

The data discussed in the previous paragraph indicate that IL-10 suppresses bone formation by altering the osteogenic differentiation of uncommitted mesenchymal cells via an indirect pathway, rather than by directly blocking bone protein synthesis by mature bone cells. This is not surprising since IL-10 activity in other biological systems can, in large part, be explained by indirect pathways involving its suppressive effect on cytokine synthesis. Among the cytokines and growth factors which modulate bone formation and differentiation, especially the members of the TGF-β family have a vast record of well defined effects.[18,19] Apart from controlling the proliferation and differentiation of several cell types specific to bone including mesenchymal cells, chondrocytes, osteoblasts and osteoclasts.[20] In addition, TGF-β is secreted by osteoblasts and determines their phenotype in vitro.[21] In vivo, TGF-β is present in large amounts in the bone matrix where it is believed to serve as a reservoir that controls the osteogenic differentiation of "naive" mesenchymal cells.[22,23] This is illustrated by the fact that in the presence of anti-TGF-β blocking antibod-

ies bone marrow stroma cultured in the presence of β-glycerophosphate and vitamin C does not synthesize bone proteins nor does it form a mineralized matrix.[24] In contrast however, anti-TGF-β blocking antibodies do not influence the stroma proliferation and the number or size of CFU-F. Therefore, it is obvious that TGF-β plays an important role during the first steps of osteoblasic differentiation.[24] Taken together, the obvious question to be raised is: Does IL-10 mediate its suppressive effect on bone formation by blocking TGF-β synthesis? The fact that TGF-β synthesis peaks when IL-10 induces its suppressor effect is circumstantial but definitely points in the direction that this indeed is the case. IL-10 suppresses TGF-β synthesis in marrow stroma cultures as illustrated at the mRNA and protein level. In addition, the reduced ALP, collagen type I and osteocalcin synthesis and matrix mineralization in the IL-10 treated cultures could be restored by adding exogenous TGF-β.[24] Hence, it is fair to state that IL-10 suppresses bone formation by blocking the synthesis of TGF-β, which is an essential factor in the commitment process of mesenchymal cells to the osteogenic lineage.

IL-10 AND HEMATOPOIESIS

IL-10 SUPPORTS THE GROWTH OF HEMATOPOIETIC PROGENITOR CELLS

As mentioned above, IL-10 was initially identified as "cytokine synthesis inhibiting factor" (CSIF), based on its ability to suppress IL-2, IL-3, lymphotoxin, IFN-γ and GM-CSF synthesis by Type 1 helper T cells.[25,26] Further studies showed that IL-10 exerted a much broader range of biological activities including, the induction of thymocyte[27] and B cell[28] growth and the suppression of macrophage functions[29] and bone formation[16,24] The first indication of IL-10 potential to regulate hematopoiesis was based on its stimulatory effect on mast cells.[30] Whereas, IL-10 alone had only a minor stimulatory effect, it significantly stimulated the growth of mast cell progenitor cells in the presence of IL-3. Along the same line, a recent study showed that IL-10 costimulates colony formation by purified progenitor and stem cells in the presence of a variety of cytokines.[31] IL-10 induced a modest, but significant stimulation of the growth of megakaryocytes and mixed colonies in combination with either IL-3, IL-6, IL-11 or Epo. Apart from costimulating committed progenitor cells, IL-10 also co-stimulates colony formation by purified Thy1^{lo} Sca1$^+$ stem cells in combination with IL-3, IL-1, IL-6 and G-CSF. In conclusion, these studies clearly indicate that IL-10's potential to stimulate purified hematopoietic cells is determined by the presence of other cytokines. In an in vivo setting, this implies that the biological potential of IL-10 may in large part depend on cytokines synthesized by the microenvironment. In this context, we performed a series of experiments to identify the effect of IL-10 on hematopoietic cells in the presence of stromal cells. We observed that osteogenic cultures lose the ability to form an extracellular matrix and acquire a "Dexter-like" morphology in the presence of IL-10.[16,24] Hence, in contrast to normal conditions, murine osteogenic bone marrow cultures do support the growth of hematopoietic cells in the presence of IL-10. Indeed, whereas cells from osteogenic cultures are unable to generate granulocyte/macrophage colonies (CFU-GM) after 2 weeks of culture, cells from IL-10 treated cultures produced CFU-GM up to 8 weeks. It is conceivable that the long term generation of colony forming cells in the latter cultures is due to a direct effect of IL-10 on the growth of myeloid progenitor cells. Our experiments exclude this possibility since IL-10 alone did not support the production of CFU-GM from either normal bone marrow or from cells derived from untreated or IL-10 treated bone marrow cultures. Hence, IL-10 may act by stimulating the growth of self-renewing stem cells in osteogenic cultures. The fact that cells from IL-10 treated osteogenic cultures produce CFU-GM for several weeks when plated on irradiated stroma layers illustrates the presence of long term culture initiating cells (LTC-IC). These cells are considered to be hematopoietic stem cells which have the ability to repopulate the bone marrow after transplantation.[32] The presence of hematopoietic stem cells is furthermore supported by the increased numbers of c-kit$^+$ in IL-10 treated cultures. Taken together, these data illustrate that IL-10 can stimulate both the growth of progenitor and stem cells in the absence of other exogenous cytokines, provided the presence of a stromal microevironment.[33]

IL-10 STIMULATES HEMATOPOIESIS BY BLOCKING ENDOGENOUS TGF-β1 SYNTHESIS

Thus far, it is clear that IL-10 can induce the growth of CFU-GM in osteogenic cultures.[33] However, it is unclear by which mechanism this happens since IL-10 alone does not have the potential to stimulate the proliferation of myeloid progenitors. Since IL-10 activity depends on the presence of either exogenous cytokines or a stromal microenvironment, it is fair to assume that IL-10 acts via an indirect pathway. The observation that IL-10 suppresses the synthesis of TGF-β1 is of interest in this context.[24] The TGF-β family is a family of polypeptides with pleiotropic biological activities on a variety of cells including hematopoietic cells.[34] Especially TGF-β1 was shown to exert suppressive effects on stem cell growth and colony formation.[35] More precisely, TGF-β1 prevents hematopoietic stem cells from entering the cell cycle, which explains their quiescent status in vivo and in vitro.[36] In other words, TGF-β1 can be considered as a constitutive suppressor of a variety of hematopoietic and imunological functions.[37,38] This is illustrated very nicely in TGF-β1 knock-out mice, which die soon after birth due to uncontrolled, multifocal infections.[37] All these data let us suggest that IL-10 stimulatory effect on hematopoiesis may be related to its potential to block TGF-β1 synthesis.[24] It is conceivable that IL-10 acts by suppressing TGF-β1 synthesis by the stromal cells, which subsequently results in increased hematopoietic activity. This idea is supported by the observation that IL-10 treated osteogenic cultures loose their Dexter-like morphology and start to produce large amounts of bone proteins in the presence of exogenous TGF-β1. Furthermore, osteogenic cultures support hematopoiesis in the presence of anti-TGF-β1 antibodies. In combination with the data shown above, it appears that IL-10 directs stromal differentiation toward hematopoietic support and that at the same time it suppresses TGF-β1 synthesis, which acts as a suppressor of hematopoiesis.

IL-10 AND STROMA MORPHOLOGY

In vitro bone formation is characterized by a number of characteristic morphologic features.[11] Bone marrow stroma in the presence of β-glycerophosphate and vitamin C forms an adherent layer of polygonal, fibroblastoid cells (Fig. 7.2A). These cells form bone ossicles which mineralize. Scanning electron microscopy reveals that these nodules contain large amounts of collagen, which are in close contact with hydroxyapatite deposits, which is the most abundant and most stable type of calciumphosphate in the bone[11] (Fig. 7.2D). In cultures supplemented with IL-10 or anti-TGF-β this bone morphology disappears and is replaced by adherent rounded cells, which are covered by semi- to non-adherent cells (Fig. 7.2B and 7.2C, respectively). The data above clearly demonstrate that the latter cells are myeloid progenitor cells and stem cells. In addition, scanning electron microscopy shows that in the presence of IL-10 or anti-TGF-β antibodies there is no sign of a mineralized matrix[16,24] (Fig. 7.2E and 7.2F). In general, it can be stated that these cultures have a lot in common with hematopoietic cultures as described by Dexter.[39,40] Indeed, characteristic for these cultures is the presence of blanket stroma that is in direct proximity with maturing hematopoietic cells (Fig. 7.2E and 7.2F). Apart from these eye-striking effects, IL-10 treated cultures contain also a number of characteristic cell types. One of them are multinucleated cells which stain for tartrate resistant acid phosphatase (Fig. 7.2H). The number of these cells is significantly increased in IL-10 treated cultures (Table 7.1). The presence of this enzyme in murine cells refers to their osteoclastic characteristic. This fits with the observation that IL-10 treated cultures contain large numbers of macrophages.[33]

Consequently, one additional biological effect of IL-10 may be the stimulation of macrophage fusion leading to the formation of osteoclasts. At this point this possibility has to be explored further. On the other hand, IL-10 treated cultures show increased numbers of "dendritic cells" (Fig. 7.2I). At this point, this nomenclature refers to the morphology of these cells rather than to their function. In any case, since these "octopus-like" cells connect cells in the culture, their function may be associated with cell-cell contact and cell signaling between hematopoietic cells. Another abundant cell type in the IL-10 treated cultures are fat cells (Fig. 7.2G). This observation fits in IL-10 stimulatory activity on the hematopoietic system since

Table 7.1. Effect of IL-10 on the generation of TRAP positive, multinucleated cells in osteogenic cultures of mouse bone marrow

	Number of cells per culture	
	Control	IL-10*
TRAP+	1.5 ± 1.9	192 + 15
TRAP–	133 ± 25	64 + 19

* IL-10 was added at 5x10³ U/ml

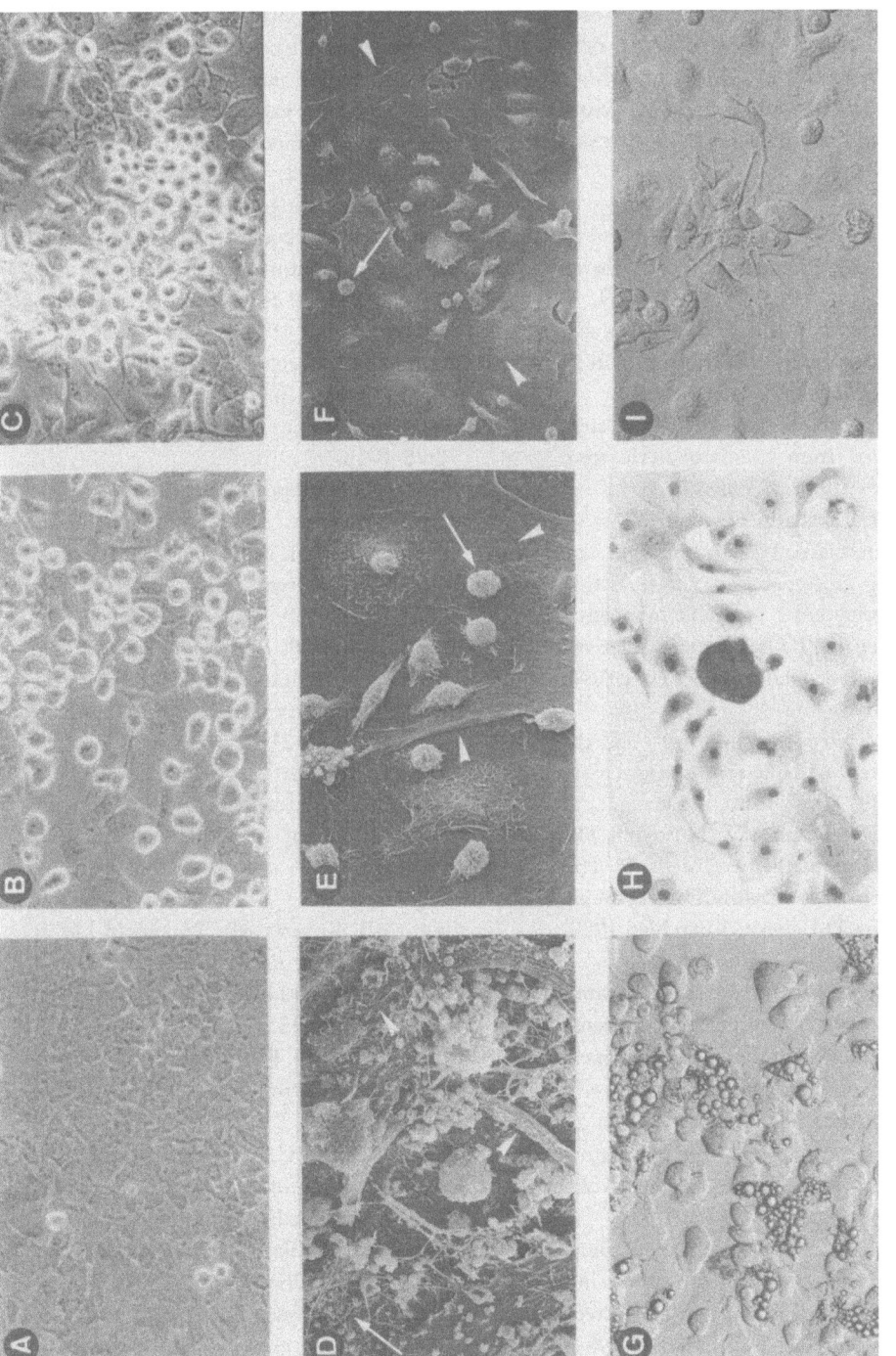

Fig. 7.2. Effect of IL-10 and anti-TGF-β antibodies on the morphology of bone marrow stroma cultures. Light microscopy (phase contrast, x320) image of 15-day old control (A), IL-10 treated (B) and anti-TGF-β treated (C) bone marrow cultures. Flat adherent polygonal cells characterize the control cultures. Clusters of semi-and non-adherent cells in contact with reticular and fibroblastic cells characterize the IL-10 and anti-TGF-β treated cultures. Scanning electron microscopy of 25 day old control (D, x900), IL-10 treated (E, x700) and anti-TGF-β treated (F, x600) cultures. Note the mineralized collagenous matrix (white arrow) surrounding fibroblastic cells (white arrow heads) in the control culture. IL-10 and anti-TGF-β treated cultures showed no mineralized matrix and contain myeloid cells (white arrow) and blanket cells (white arrow heads). Hoffman contrast micrograph of a 15 day old, IL-10 treated culture showing adipocytes (G, x320). Osteoclastic, TRAP+ cell in a 20 day old, IL-10 treated culture (H, x320). Dendrite-like cell in a 20 day old, IL-10 treated culture (I, x320).

fat cells have been previously associated with hematopoiesis.[41] Indeed, it is well documented that fat cells are a necessary constituent in Dexter cultures.[39,40] Although there is no direct proof, it is conceivable that the occurrence of these cells results from the reduced TGF-β synthesis in the IL-10 treated cultures. A previous study illustrated that TGF-β suppresses adipocyte formation.[42]

REFERENCES

1. Friedenstein AJ. Precursor cells of mechanocytes. Int Rev Cyto 1976; 47: 327-59.
2. Owen ME. Lineage of osteogenic cells and their relationship to the stromal system. Calcif Tissue Int 1994; 36:S5-S6.
3. Ash P, Loutit JF, Townsend KMS. Osteoclasts derived from haematopoietic stem cells. Nature 1980; 283:669-70.
4. Udagawa N, Takahashi N, Akatsu T et al. Origin of osteoclasts: Mature monocytes and macrophages are capable of differentiating into osteoclasts under a suitable microenvironment prepared by bone marrow-derived stromal cells. Proc Natl Acad Sci USA 1990; 87:7260-64.
5. Beresford J. Osteogenic stem cells and the stromal system of bone and marrow. Clin Orthop 1989; 240:270-80.
6. Fiorentino DF, Bond MW, Mosmann TR. Two types of mouse helper T cells. V. Th2 clones secrete a factor that inhibits cytokine production by Th1 clones. J Exp Med 1989; 170:2081-95.
7. Friedenstein A, Chailaklyan RK, Latsinik NV et al. Stromal cells responsible for transferring the microenvironment of hemopoietic tissue Transplantation 1974; 17:331-40.
8. Ashton BA, Allen TD, Howlett CR, et al. Formation of bone and cartilage by marrow stromal cells in diffusion chambers in vivo. Clin Orthop 1980; 151:294-307.
9. Benayahu D, Kletter D, Zipori D et al. Bone marrow derived stromal cell line expressing osteoblastic phenotype in vitro and osteogenic activity in vivo. J Cell Physiol 1989; 140:1-7.
10. Mathieu E, Schoeters G, Van der Plaetse R et al. Establishment of an osteogenic cell line derived from adult mouse bone marrow stroma by use of a recombinant retrovirus. Calcif Tissue Int 1992; 50:362-71.
11. Falla N, Van Vlasselaer P, Bierkens J et al. Characterization of a 5-Fuorouracil enriched osteoprogenitor population of the murine bone marrow Blood 1993; 82:3580-91.
12. Jacobsen K, Miyake K, Kincade PW et al. Higly restricted expression of a stromal cell determinant in mouse bone marrow in vivo. J Exp Med 1992; 176:927-35.
13. Imhof BA, Schlinger C, Handloser K et al. Monoclonal antibodies that block adhesion of B cell progenitors to bone marrow stroma in vitro prevent B cell differentiation in vivo. Eur J Immunol 1991; 21:2043-49.
14. Simmons PJ, Torok-Storb B. CD34 expression by stromal precursors in normal human adult bone marrow. Blood 1991; 78:2848-53.
15. Huang S, Terstappen LWMM. Formation of haemopoeietic micro-environment and hemopoietic stem cells from single human bone marrow stem cells. Nature 1993; 360:745-49.
16. Van Vlasselaer P, Borremans B, Van Den Heuvel R et al. Interleukin 10 inhibits the osteogenic activity of mouse bone marrow. Blood 1993; 82:2361-70.
17. Sudo H, Kodama H, Amagai Y et al. In vitro differentiation and calcification in a new clonal osteogenic cell line from newborn mouse calvaria. J Cell Biol 1983; 96:191-98.
18. Sporn MB, Roberts AB, Wakefield LM et al. Transforming growth factor-β: biological and biochemical function and structure. Science 1986; 233:532-34.
19. Massague J. The TGF-β family of growth and differentiation factors. Cell 1987; 49:437-38.
20. Noda M, Rodan GA. Type-beta transforming growth factor inhibibts proliferation and expression of alkaline phosphatase in murine oosteoblast-like cells. Biochem Biophys Res Commun 1986; 140:56-65.
21. Rosen DM, Stempien SA, Thompson AY et al. Transforming growth factor-beta modulates the expression of osteoblast and chondroblast phenotypes in vitro. J Cell Physiol 1988; 134:337-46.
22. Hauschka PV, Maurakos AE, Lafraty MD

et al. Growth factors in bone matrix. Isolation of multiple types by affinity chromatography on heparin-Sepharose. J Biol Chem 1986; 261:12665-74.

23. Gehron-Robey PC, Young MF, Flanders KC et al. Osteoblasts synthesize and respond to TGF-β in vitro. J Cell Biol 1987; 105: 457-63.

24. Van Vlasselaer P, Borremans B, Van Gorp U et al. Interleukin 10 inhibits transforming growth factor-β synthesis required for osteogenic commitment of mouse bone marrow cells. J Cell Biol 1994; 124:569-77.

25. Fiorentino DF, Zlotnik A, Mosmann TR et al. IL-10 inhibibts cytokine production by activated macrophages J Immunol 1991; 147:3815-22.

26. de Waal Malefyt R, Abrams J, Bennett B et al. IL-10 inhibibts cytokine synthesis by human monocytes: an autoregulatory role of IL-10 produced by monocytes.. J Exp Med 1991; 174:1209-20.

27. MacNeil IA, Suda T, Moore K et al. Interleukin 10: A novel growth co-factor for mature and immature thymocytes. J Immunol 1990; 145:4167-73.

28. Go N, Castle B, Barrett R et al. Interleukin 10; A novel B cell stimulatory factor. Unresponsiveness of X chromosome-linked immunodeficiency B cells. J Exp Med 1990; 172:1625-31.

29. de Waal Malefyt R, Haanen J, Spits H et al. IL-10 and viral IL-10 strongly reduce antigen-specific human T cell proliferation by diminishing the antigen presenting capacity of monocytes via downregulation of Class II MHC expression. J Exp Med 1991; 174:915-24.

30. Thompson-Snipes L, Dhar V, Bond MW et al. Interleukin 10: A novel stimulatory factor for mast cells and their progenitors. J Exp Med 1991; 173:507-10.

31. Rennick D, Hunte B, Dang W et al. Interleukin 10 promotes the growth of megakaryocytes, mast cells, and multilineage colonies; analysis with committed progenitors and Thy1loSca1+ stem cells. Exp Hematol 1994; 22:136-41.

32. Ploemacher RE, van der Sluijs JP, Voerman JSA et al. An in vitro limiting dilution assay of long-term repopulating hematopoietic stem cells in the mouse. Blood 1989; 74:2755-63.

33. van Vlasselaer P, Falla N, van den Heuvel R, Dasch J, de Waal Malefijt R. Interleukin-10 (IL-10) drives osteogenic bone marow stroma towards hematopoietic support by blocking endogenous transforming growth factor beta (TGF-β) synthesis. Clin Ortho and Rel Res 1994.

34. Ohta M, Greenberger JS, Anklesaria P et al. Two forms of transforming growth factor β distinguished by multipotential hematopoietic progenitor cells. Nature 1987; 329:539-41.

35. Carlino JA, Higley HR, Creson JR et al. Transforming growth factor β1 systemically modulates granuloid, erythroid, lymphoid and thrombocytic cells in mice. Exp Hematol 1992; 20:943-50.

36. Hatzfeld J, Li ML, Brown EL et al. Release of early hematopoietic progenitors from quiescence by antisense transforming growth factor β1 or Rb oligonucleotides. J Exp Med 1991; 174:925-29.

37. Shull MM, Ormsby I, Kier AB et al. Targeted disruption of the mouse transforming growth factor β1 gne results in multifocal inflammatory disease. Nature 1992; 359: 693-99.

38. Hayashi SI, Gimble JM, Henley A et al. Differential effects of TGF-β1 on lympho-hemopoiesis in long-term bone marrow cultures. Blood 1989; 74:1711-19.

39. Dexter T. Stromal cell associated hematopoiesis. J Cell Physiol 1982; (suppl 1):87-94

40. Allen TD, Dexter T. The essential cells of the hematopoietic environment. Exp Hematol 1984; 12:517-21.

41. Tavassoli M. Marrow adipose cells and hematopoiesis: an interpretative review. Exp Hematol 1984; 12:139-46.

42. Ignotz RA, Massague J. Type β transforming growth factor controls the adipogenic differentiation of 3T3 fibroblasts. Proc Natl Acad Sci USA 1985; 82:853034.

43. van Vlasselaer P, Falla N, Snoeck H, Mathieu E. Characterization and purification of osteogenic cells from murine bone marrow by two-color cell sorting using anti-Sca-1 monoclonal antibody and wheat germ agglutinin. Blood 1994; 84:753-63.

REGULATORY FUNCTION OF IL-10 IN EXPERIMENTAL PARASITIC INFECTIONS

Steven G. Reed and Alan Sher

INTRODUCTION

O ur understanding of the cellular and molecular nature of the immune response to infectious pathogens is currently expanding at an
unprecedented pace. To a large extent, this is due to advances in the
area of cytokine biology. Studies on the regulation of cytokines during
infection have shown these molecules to be potent mediators of both
resistance and disease. The presence or absence of a particular cytokine
can be a predictable indicator of disease outcome. Cytokine patterns can
now be used to investigate mechanisms of pathogen virulence. They can
explain why certain infectious organisms, particularly parasites, cause a
generalized depression of the host immune system, and can also be ex-
ploited to design novel intervention strategies unthinkable a few years
ago. Because parasites are often adapted to long-term survival in their
mammalian hosts, and thus cause chronic infections, they are in many
cases exceptional systems in which to study immunoregulatory mechanisms. Studies on one of the most potent regulatory cytokines, IL-10,
have been performed in a variety of infectious disease systems. Results
from studies of two parasitic infections, *Trypanosoma cruzi* and *Schistosoma
mansoni*, are discussed in this chapter.

REGULATION BY IL-10 OF *T. CRUZI* INFECTION

The intracellular protozoan parasite *T. cruzi* replicates in mammalian
hosts within the cytoplasm of several types of nucleated cells, including
macrophages. The disease caused by *T. cruzi* infection is manifested as
acute or chronic. Acute infections are characterized by rapid parasite mul-

Interleukin-10, edited by Jan E. deVries and René de Waal Malefyt.

tiplication, whereas chronic infection involves relatively low but persistent numbers of parasites in host tissue. Recent studies using mice genetically susceptible or resistant to acute *T. cruzi* infection have been useful for demonstrating a key role for macrophage inactivating cytokines in determining the infection course. Two cytokines which have been shown to be critical in determining whether disease will develop are transforming growth factor-beta (TGF-β) and IL-10, the subject of this chapter. In vitro, IL-10 has been shown to be an effective antagonist of parasite killing by activated macrophages.[1] The ability of IL-10 to block the IFN-γ-mediated killing of intracellular *T. cruzi* was found to correlate with decreased nitrite generation.[2] A similar finding has been made with another protozoan parasite, *Toxoplasma*.[3] The ability of IL-10 to inhibit the effects of IFN-γ has particularly important implications for these intracellular protozoa, resistance to which is heavily dependent on IFN-γ. The following discussion will highlight in vivo results regarding the role of IL-10 in susceptibility to *T. cruzi* infection.

The course of *T. cruzi* infection in different mouse strains is influenced by both the inoculum number and parasite isolate,

and can range from mild to fatal outcomes. A cloned *T. cruzi* of the Tulahuen strain was used to study mechanisms of disease resistance or susceptibility. C57BL6 (B6) mice are highly susceptible and develop acute infections during a course of 18-25 days. Hybrid C57BL6 x DBA2 (B6D2) mice are highly resistant and develop nonfatal infections with very low parasitemias and no visible morbidity. The immunological basis for genetic resistance or susceptibility is a subject of particular interest, and one that can be addressed with this model system. One aspect of *T. cruzi* infections in susceptible mouse strains is the resulting decrease in immune responsiveness. Because IL-10 has been associated with decreased cytokine production[4] and decreased macrophage activation,[5] we compared the production of biologically active IL-10 by susceptible B6 and resistant B6D2 mice during the course of acute infection with *T. cruzi*. While no IL-10 production was seen in B6D2 mice, significant levels of IL-10, as determined by the ability of supernatants from spleen cell cultures to inhibit the production of IFN-γ, were produced by infected B6 mice (Fig. 8.1). This result did not address whether IL-10 production was the cause or effect of increased disease. However, it did provide an

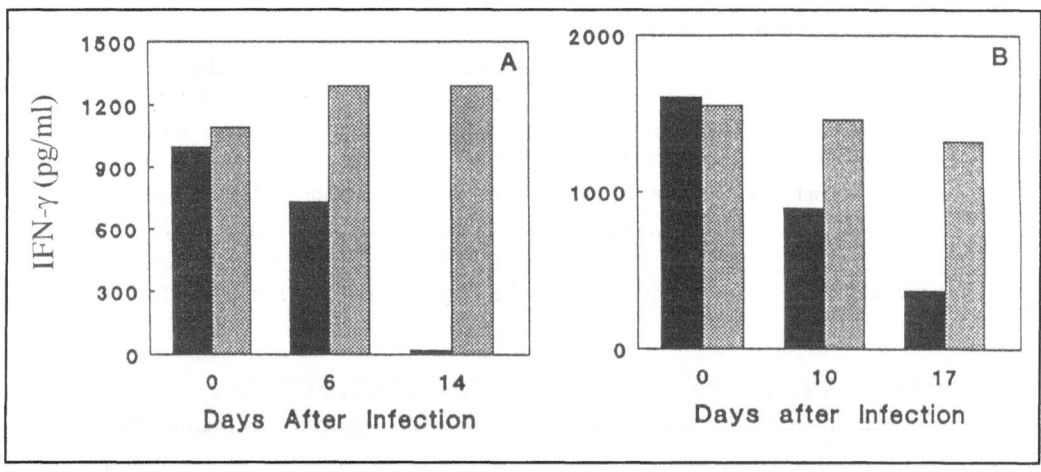

Fig. 8.1. *IL-10 activity in spleen cell supernatants from B6 (filled bar) and B6D2 (stippled bar) mice infected with* T. cruzi. *48-h supernatants of 5 x 10⁶ unstimulated spleen cells from infected and from mice infected for 6-17 d with* T. cruzi *were assayed for CSIF activity. Representative data from two experiments are shown. Each point represents pooled spleen cells from three mice. Reprinted with permission from Rockefeller University Press. J Exp Med 1992; 175:169-174.*

association between the production of active IL-10 and disease susceptibility.

IL-10 can inhibit both the production of IFN-γ by T cells, as well as the utilization of IFN-γ by macrophages. IFN-γ is an important effector cytokine for the inhibition of *T. cruzi* in vitro and in vivo.[6,7] We therefore compared the kinetics of the in vivo production of both IFN-γ and IL-10 in resistant and susceptible mice. Confirming the findings on IL-10 biological activity, freshly harvested spleen cells from susceptible but not resistant mice, had IL-10 mRNA by 8 days or longer after infection. In contrast, mRNA for IFN-γ was detected in both resistant and susceptible strains. These observations supported the concept that IL-10 production in response to infection was closely correlated with disease susceptibility. The cell source of the IL-10 production has been addressed in a recent study (Reed et al, manuscript submitted). Our results demonstrate that macrophages, B cells, and T cells can all produce IL-10 in susceptible mice. In addition, experiments performed in SCID mice also demonstrated that macrophages were an important source of IL-10 production during *T. cruzi* infection.

The role of IFN-γ as an effector mechanism in protection against disease from *T. cruzi* infection has been indicated by the demonstration that exogenous IFN-γ can protect susceptible mice,[7] and that in vivo neutralization of endogenous IFN-γ leads to the development of fatal infections in genetically resistant mice.[1,8] As shown in Fig. 8.2, B6D2 mice treated with anti-IFN-γ (Xmg 1.2) developed fatal infections, resembling those in susceptible B6 mice. The development of disease following anti-IFN-γ treatment was paralleled by the appearance of IL-10 mRNA in the spleen of the treated mice. Whereas infected B6D2 mice treated with saline or control mAb had no detectable IL-10 during the course of infection, mice treated with anti-IFN-γ had detectable IL-10 mRNA by day 6 after infection, resembling the pattern seen in infected B6 mice. Thus, in vivo neutralization of IFN-γ in resistant mice produced a disease susceptible phenotype characterized by increased IL-10 production and fatal infections. It is likely that macrophages are an important, if not the primary source of IL-10 in anti-IFN-γ-treated B6D2 mice. Support for this came from experiments in which SCID mice were treated with anti-asialo GM-1 antibody prior to infection with *T. cruzi*. This treatment effectively eliminated NK cell activity and IFN-γ production, while correspondingly increasing IL-10 mRNA in peritoneal cells compared to control Ab treated SCID mice. Together, the in vivo depletion results obtained in B6D2 and SCID mice suggest that IFN-γ production during *T. cruzi* infection is an important mechanism for down regulating IL-10 production and disease susceptibility.

Although the above results have suggested a close correlation between IL-10 production and susceptibility to *T. cruzi* infection, more definitive information was obtained from experiments involving in vivo neutralization of endogenous IL-10. Two different anti-Mu-IL-10 mAb were used in these experiments, both obtained from DNAX. One was SXC-1, an IgM, and the other was JES5-2A5, a rat IgG 2A. To test the ability of these mAb to alter the course of *T. cruzi* infection, they were injected separately into susceptible B6 or BALB/c mice, followed by infection with a lethal inoculum of *T. cruzi*.

In these experiments (Reed et al, J Immunol, in press), it was found that treatment with anti-IL-10 mAb prevented the development of acute disease and death in susceptible B6 mice. The treatment was most effective if given prior to infection. Because the production of IL-10 is a relatively early event following infection, it is important that an effective neutralization therapy be given early.

The results from studies on experimental *T. cruzi* infection have provided some potentially important insights into molecular mechanisms of disease susceptibility following infection. It is becoming evident that important macrophage cytokines, such as IL-10, play an early and essential role

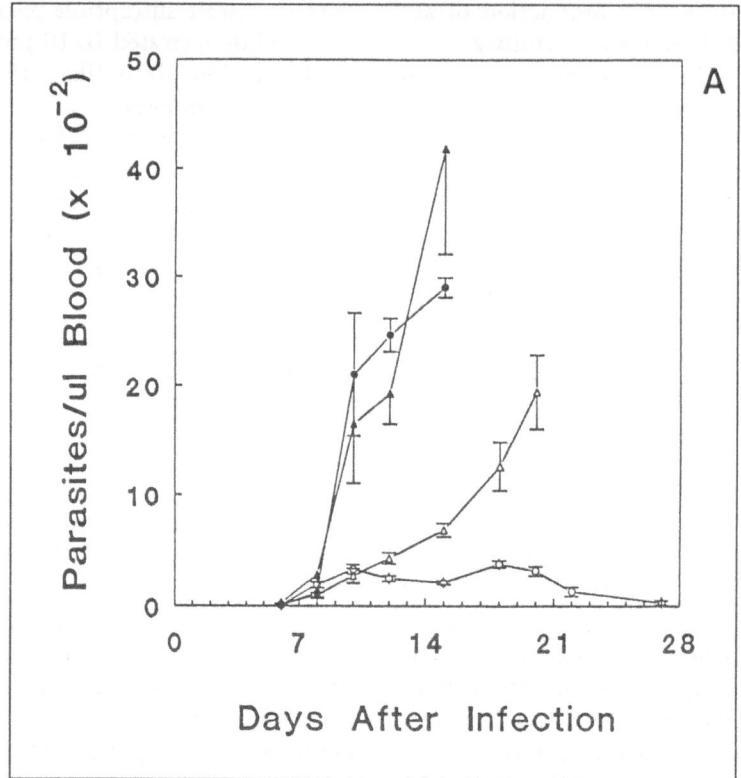

Figs. 8.2 A&B. Effects of anti-IFN-γ treatment in resistant B6D2 (circles) and susceptible B6 (triangles) mice infected with T. cruzi. Mice were treated with 1 mg XMG 1.2 (closed symbols) or saline (open symbols) 1 d before and 3 d after intraperitoneal infection with 10³ trypomastigotes. Parasitemias (A) and mortality (B) were determined at indicated days after infection. Each point (mean ± SEM) represents six mice. Mice treated with XMG 1.2 had parasitemias significantly higher (p < 0.05) than control mice at day 10 and beyond. In B, no mortality was seen in saline-treated B6D2 mice, as compared with 100% mortality in XMG-treated (B6D2 XMG) mice. Reprinted with permission from Rockefeller University Press. J Exp Med 1992; 175:169-174.

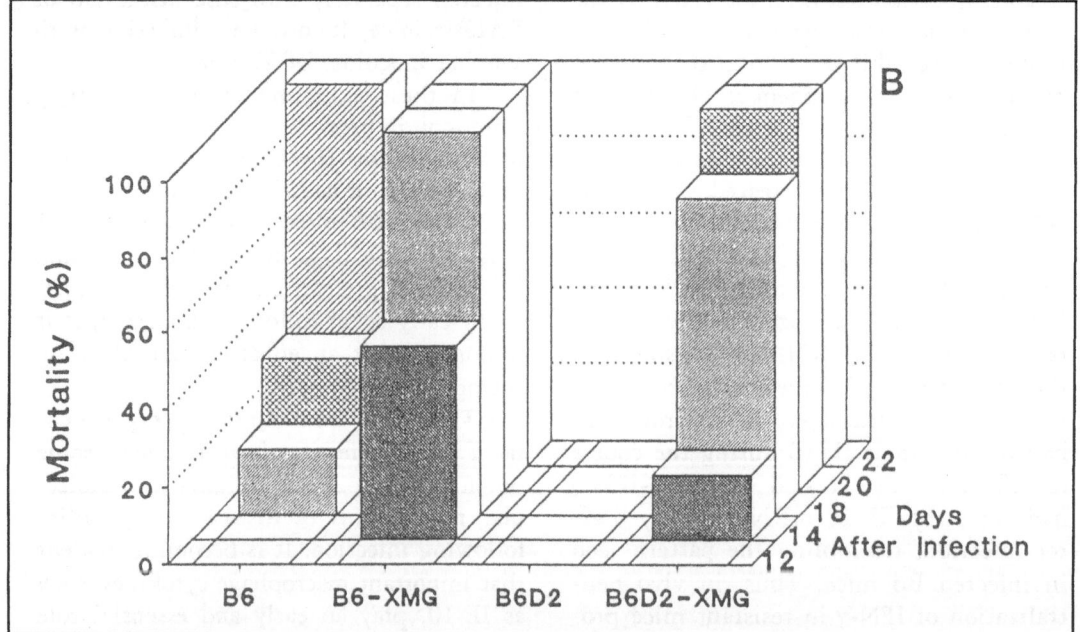

in influencing T-cell response patterns and disease outcome. Conversely, IFN-γ produced by T cells and NK cells can influence the production of IL-10. Experimental *T. cruzi* infections have already demonstrated that disease susceptibility is not related to the absence of IFN-γ production, as has been suggested in experimental leishmaniasis.[9] Rather, the presence of IL-10 appears to be at least one of the important cytokine determinants of disease susceptibility. We recently made similar observations in human visceral leishmaniasis. In these studies, cytokine patterns from patients during the acute vs. convalescent phase differed most clearly in their production of IL-10. For example, although IFN-γ mRNA was present in lymphoid tissue during both stages, mRNA for IL-10 was found only during acute disease.[10,11] Like experimental *T. cruzi* infections, human visceral leishmaniasis is an acute disease which if not treated, is often fatal. Whether increased IL-10 production is a cause or effect of increased disease susceptibility remains to be determined, but for now, a definite association has been made.

THE ROLE OF IL-10 IN EXPERIMENTAL SCHISTOSOMIASIS

A second parasite disease model where IL-10 has been implicated as playing an important immunoregulatory role is murine schistosomiasis. *Schistosoma mansoni* is a parasitic helminth which infects millions of people in developing tropical countries and which is a major cause of morbidity. Schistosome-infected humans and laboratory mice display both elevated IgE levels and eosinophilia, two of the immunologic hallmarks of worm infestations known to be controlled by the Th2 cytokines IL-4 and IL-5, respectively.[12] In the mouse model, infected animals show enhanced Th2 cytokine (IL-4, IL-5, IL-6, IL-13) production both in response to parasite antigen in vitro and in tissues in vivo.[13,14] The major stimulus of this Th2 response in murine *Schistosomiasis mansoni* is provided by parasite eggs produced by the female worms.[13] On depo-

sition in the liver, intestine, and other organs, schistosome eggs induce tissue reactions (granulomas) which lead to the major pathologic manifestations of the disease.

The induction of Th2 responses by schistosome infections appears to be accompanied by a state of cross-regulation in which in vitro Th1 cytokine responses to both parasite antigens and mitogens are suppressed.[15] Moreover, mice carrying patent schistosome infections show suppressed in vitro Th1 responses to foreign antigens (e.g., myoglobin, vaccine virus) administered in vivo.[16,17] At later stages, infected animals show a generalized suppression of all cytokine responses, a situation which resembles an anergic state.[18]

A role for IL-10 in the induction of these immunoregulatory events was suggested by the discovery that the cytokine is expressed both in vivo and in vitro during the same period as the observed downmodulation.[19] Thus, spleen cells from patently infected mice produced high levels of the cytokine when stimulated in vitro with either parasite antigen or mitogen.[19] More importantly, antibody-mediated neutralization of IL-10 in vitro resulted in restored Th1 (IFN-γ) cytokine production in response to these stimuli[19] or to myoglobin in the case of infected mice immunized with that foreign antigen.[16] The latter findings argue that IL-10 is responsible for the suppressed Th1 responses seen in infected mice. In support of this hypothesis, IL-10 knockout mice were shown to display IFN-γ mRNA levels in infected liver which were significantly elevated with respect to the low levels seen in infected heterozygote controls (Wynn, T et al, unpublished observations). CD4 cells were shown to be the major source of the IL-10 detected in the in vitro recall experiments, suggesting a state wherein Th2 lymphocytes–by means of IL-10–cross-regulate Th1 cytokine production. As described below, macrophages and B cells were later demonstrated to be additional sources of the cytokine in infected mice.

The concept that Th1 cytokine production is downregulated in schistosome infection–rather than the development of the subset inhibited—is supported by recent experiments by Vella and Pearce.[20] These investigators found that after in vitro culture for 3 days, lymph node cells from mice injected subcutaneously with eggs or spleen cells from mice with patent infections become capable of producing IFN-γ at normal levels in response to stimulation with mitogen. One explanation of their findings is that early in culture, enough IL-10 or other cross-regulating cytokine is produced to suppress Th1 responses but that, after continued incubation in vitro, this production is decreased or inhibited. Nevertheless, IL-10 and IL-4 responses were also increased following the same incubation, arguing against this hypothesis. An alternative explanation suggested by Vella and Pearce[20] is that an accessory cell (e.g., B cell or macrophage) required for IL-10 suppression of IFN-γ production is absent from or dies during culture, thereby preventing IL-10 from exerting its down-regulatory activity. Regardless, the results demonstrate that normal development of Th1 cells occurs in helminth-infected hosts despite the elevation in Th2 response.

That the IL-10 produced during schistosome infection results in altered accessory cell function has been formally demonstrated by Stadecker and colleagues. These investigators had shown that macrophages isolated from schistosome egg granulomas caused antigen-specific, MHC-restricted nonresponsiveness in Th1 cells in vitro.[21] Moreover, a similar loss in antigen-presenting cell (APC) function could be generated by incubating splenic macrophages with supernatants of egg antigen-stimulated spleen cells from schistosome-infected mice. Antibody neutralization studies demonstrated that IL-10 was the supernatant factor responsible for the inhibition in APC activity.[22] As expected, the effects of IL-10 on APC activity were restricted to Th1 clones which, in later experiments, were shown to be blocked in their ability to generate DTH responses to egg antigens in vivo. These authors speculate that the production of IL-10 and its interaction with APC results in a state of

Th1 cell anergy, which could explain the down-modulation in granuloma formation observed late in schistosome infection.[22]

In all of the experiments cited above, Th2 cells were the presumed source of the IL-10 stimulated by schistosome infection. Recent experiments by Harn and his colleagues point to an alternative cellular source of the cytokine.[23] These investigators showed that an oligosaccharide, LNFP-III, associated with schistosome egg antigens, stimulates T-depleted B lymphocyte-enriched spleen cell populations from infected mice to produce IL-10 but not IL-4. Based on the known inability of immunoglobulin crosslinking on B cells to stimulate IL-10 production, the authors argued that the observed responses are probably the result of interaction with an additional receptor such as a selectin.[23] They argue that this "non-specific" stimulation of IL-10 from expanded B cells could play a major role in Th1 cell regulation.

The existence of yet a third source of IL-10 in schistosome-infected mice was suggested by O'Garra and colleagues.[24] These investigators observed spontaneous production of IL-10 mRNA by peritoneal cells from mice early (5 wk) in infection before antigen-induced production of the cytokine by CD4⁺ T cells is observed in spleen. Ly-1 B cells purified from these peritoneal populations failed to express significant IL-10 mRNA levels, suggesting that the responding cells were non-B as well as non-T in origin and possibly macrophages; however, the identity of the IL-10-producing cells was not further characterized. In more recent work, Stadecker and colleagues (personal communication) have shown that macrophages isolated from egg granulomas secrete IL-10 in vitro, thus formally demonstrating this cellular source of the cytokine. Interestingly, the same granuloma derived macrophages show down-regulated B7 expression, a property partially reversed by anti-IL-10 treatment. Thus, as proposed by these investigators, a major activity of IL-10 in murine schistosomiasis may be the regulation of macrophage costimulatory function.

It is clear from the studies summarized above that IL-10 can be produced by a number of different cell types during schistosome infections and has multiple immunoregulatory activities; nevertheless, the key question raised by the association of IL-10 with schistosomiasis concerns the possible functional role(s) of the cytokine in the biology of schistosome infection. In other words, if the major activities of IL-10 are downregulatory, why would schistosomes evolve a relationship with the immune system which favors the induction of this cytokine? A possible explanation comes from studies on the effects of IL-10 on cell-mediated immunity against schistosomula, the early larval stage of the parasite. Previous work had indicated that IFN-γ-activated macrophages are potential killers of schistosomula in vitro and had implicated this mechanism in the resistance to infection induced by attenuated vaccination.[25] Nitrogen oxides (NO) were shown to be the principal biochemical agent responsible for parasite killing mediated by these effectors.[26] Experiments by Oswald, Gazzinelli, and colleagues demonstrated that IL-10 inhibits activation of thioglycollate elicited macrophages, preventing both NO production and schistosomula killing;[3] moreover, IL-10 was shown to be even more potent in suppressing macrophage effector function in the presence of small amounts of IL-2 and TGF-β, two other downregulatory cytokines produced during schistosome infection.[27] In further experiments exploring the mechanism of IL-10 inhibition, IL-10 was shown to mediate its suppression of NO production by blocking the production of TNF-α, a monokine required as a costimulatory signal for triggering IFN-γ-exposed macrophages.[28]

The studies summarized above indicate that IL-10, in addition to inhibiting the production of IFN-γ by Th1 cells, can block the ability of this lymphokine to activate macrophages for parasite killing. Thus, the stimulation of IL-10 production can be seen as a mechanism whereby the parasite protects itself against the potentially lethal effects of the host cell-mediated immune response.[29] Such a strategy—which is reminiscent of the production of a mimicked IL-10 molecule by Epstein-Barr virus[30] would make good evolutionary sense for schistosomes which dwell during their development in the vertebrate host in both tissues and blood and, thus, require mechanisms for evading lymphokine-activated macrophages and other effector cells to which they are directly exposed.[18]

CONCLUDING REMARKS

It is evident from these and other studies discussed in accompanying chapters that IL-10 has emerged as a pivotal cytokine in the regulation of parasitic infections. Over-production of IL-10 can have detrimental effects on host resistance to both protozoal and helminth pathogens, in part by inhibiting the effects of IFN-γ. This has been clearly shown by the reversal of disease progression using an IL-10 antagonist. A similar result has been reported in experimental fungal infections.[31] Our understanding of pathogenesis of infectious agents has been greatly expanded by the discovery of IL-10, and new therapeutic approaches may result from this understanding. Elucidation of immunoregulatory events offers new hope for the prevention and treatment of parasites, among the most widespread and neglected of human infections.

REFERENCES

1. Silva JS, Morrissey PJ, Grabstein KH et al. Interleukin-10 and interferon gamma regulation of experimental *Trypanosoma cruzi* infection. J Exp Med 1992; 175:169-174.
2. Gazzinelli RT, Oswald IP, Hieny S et al. The microbicidal activity of interferon-γ treated macrophages against *Trypanosoma cruzi* involves an L-arginine-dependent, nitrogen oxide-mediated mechanism inhibitable by interleukin-10 and transforming growth factor-β. Eur J Immunol 1992; 22:2501-2506.
3. Gazzinelli RT, Oswald IP, James SL et al. IL-10 inhibits parasite killing and nitrogen oxide production by IFN-γ activated macrophages. J Immunol 1992; 148:1792-1796.

4. Fiorentino DF, Bond MW, Mosmann TR. Two types of mouse helper T cell. IV. Th2 clones secrete a factor that inhibits cytokine production by Th1 clones. J Exp Med 1989; 170:2081-2089.

5. Bogdan C., Vodovotz Y, Nathan C. Macrophage deactivation by Interleukin-10. J Exp Med 1991; 174:1549-1557.

6. Reed SG, Nathan CF, Pihl DL et al. Recombinant granulocyte-macrophage colony-stimulating factor activates macrophages to inhibit *Trypanosoma cruzi* and release hydrogen peroxide. Comparison to interferon-gamma. J Exp Med 1987; 166:1734-1746.

7. Reed, SG. In vivo administration of recombinant IFN-gamma induces macrophage activation, and prevents acute disease, immune suppression, and death in experimental *Trypanosoma cruzi* infections. J Immunol 1988; 140:4342-4347.

8. Torrico F, Heremans H, Rivera MT et al. Endogenous IFN-γ is required for resistance to acute *Trypanosoma cruzi* infection in mice. J Immunol 1991; 146:3626-3632.

9. Heinzel FP, Sadick MD, Holaday BJ et al. Reciprocal expression of interferon γ or interleukin 4 during the resolution of progression of murine leishmaniasis. Evidence for expansion of distinct helper T cell subsets. J Exp Med1989; 169:59-72.

10. Ghalib HW, Piuvezam MR, Skeiky YAW et al. Interleukin-10 production correlates with pathology in human *Leishmania donovani* infections. J Clin Invest 1993; 92:324-329.

11. Karp CL, El-Saji S.H., Wynn TA et al. In vitro cytokine profiles in patients with Kala-azar. Marked elevation of both Interleukin-10 and interferon-γ. J Clin Invest 1993; 91:16441648.

12. Finkelman FD, Pearce EJ, Urban JF Jr et al. Regulation and biological function of helminth-induced cytokine responses. In: Ash C, Gallagher RB, eds. Immunology Today. Cambridge: Elsevier Trends Journals 1991; A62-A66.

13. Grzych JM, Pearce E, Cheever A et al. Egg deposition is the major stimulus for the production of Th2 cytokines in murine schistosomiasis mansoni. J Immunol 1991;146:1322-1327.

14. Wynn T, Eltoum L, Cheever AW et al. Analysis of cytokine mRNA expression during primary granuloma formation induced by eggs of *Schistosoma mansoni*. J Immunol 1993;151:1430-1440.

15. Pearce EJ, Caspar P, Grzych JM et al. Down-regulation of Th1 cytokine production accompanies induction of Th2 responses by a parasitic helminth, *Schistosoma mansoni*. J Exp Med 1991; 173:159-166.

16. Kullberg MC, Pearce EJ, Hieny SF et al. Infection with *Schistosoma mansoni* alters Th1/Th2 cytokine responses to a non-parasite antigen. J Immunol 1992; 148:3264-3270.

17. Actor JK, Shirai M, Kullberg MC et al. Helminth infection results in decreased virus-specific CD8+ cytotoxic T cell and Th1 cytokine responses as well as delayed virus clearance. Proc Natl Acad Sci USA 1993; 90:948-952.

18. Newport G, Colley DG. Schistosomiasis. In: Warren K, ed. Immunology and Molecular Biology of Parasitic Infections. Boston: Blackwell Scientific 1993; 387-437.

19. Sher A, Fiorentino D, Caspar P et al. Production of IL-10 by CD4+ T lymphocytes correlates with down-regulation of Th1 cytokine synthesis in helminth infection. J Immunol 1991; 147:2713-2716.

20. Vella AT, Pearce EJ. *Schistosoma mansoni* egg-primed Th0 and Th2 cells: Failure to down-regulate IFN-γ production following *in vitro* culture. Scand J Immunol 1994; 39:12-18.

21. Stadecker MJ, Kamisato JK, Chikunguwo SM. Induction of T helper cell unresponsiveness to antigen by macrophages from schistosomal egg granulomas: a basis for immunomodulation in schistosomiasis? J Immunol 1990; 145:2697-2705.

22. Villanueva POF, Chikunguwo SM, Harris TS et al. Role of IL-10 on antigen-presenting cell function for schistosomal egg-specific monoclonal T helper cell responses in vitro and in vivo. J Immunol 1993; 151:3192-3198.

23. Velupillai P, Harn DA. Oligosaccharide-specific induction of interleukin-10 production by B220+ cells from schistosome-infected mice: a mechanism for regulation of CD4+ T cell subsets. Proc Natl Acad Sci USA 1994; 91:18-22.

24. Murphy E, Hieny S, Sher A et al. Detection of in vivo expression of interleukin-10 using a semiquantitative polymerase chain reaction method in *Schistosoma mansoni*-infected mice. J Immunol Meth 1993; 162:211-223.

25. James SL, Sher A. Cell-mediated immune response to schistosomiasis. Curr Top Microbiol Immunol 1990; 5:21-31.

26. James SL. The effector function of nitrogen oxides in host defense against parasites. Exp Parasitol 1991; 73:223-226.

27. Oswald IP, Gazzinelli RT, Sher A et al. IL-10 synergizes with IL-4 and transforming growth factor-β to inhibit macrophage cytotoxic activity. J Immunol 1992; 148: 3578-3582.

28. Oswald IP, Wynn TA, Sher A et al. IL-10 inhibits macrophage microbicidal activity by blocking the endogenous production of TNF-α required as a costimulatory factor for IFN-γ-induced activation. Proc Natl Acad Sci USA 1992; 89:8676-8680.

29. Sher A, Gazzinelli RT, Oswald IP et al. Role of T cell-derived cytokines in the down regulation of immune responses in parasitic and retroviral infection. Immunol Rev 1992; 127:183-204.

30. Moore KW, Vieira P, Fiorentino DF et al. Homology of cytokine synthesis inhibitory factor (IL-10) to the Epstein-Barr virus gene BCRFI. Science 1990; 248:1230-1234.

31. Romani L, Puccetti P, Mencacci A et al. Neutralization of IL-10 up-regulates nitric oxide production and protects susceptible mice from challenge with *Candida albicans*. J Immunol 1994; 152:3514-3521.

CHAPTER 9

IL-10 IN MYCOBACTERIAL INFECTION

Peter A. Sieling and Robert L. Modlin

INTRODUCTION

Infection by *Mycobacterium leprae* elicits a spectrum of clinical presentations.[1] Because these clinical manifestations correlated with the level of cell-mediated immunity (CMI) to the pathogen, leprosy serves as a useful model to study human immune responses to infection. At one pole, tuberculoid patients are able to restrict growth of the pathogen as demonstrated by the presence of relatively few skin lesions containing few bacilli. These patients have strong CMI to the pathogen as evidenced by vigorous T-cell responses in vitro and skin test reactivity to challenge with M. *leprae*. At the opposite pole, lepromatous patients have disseminated disease as manifested by numerous skin lesions containing large numbers of bacilli. The T cells of these patients are specifically unresponsive to M. *leprae* although they produce significant quantities of antibodies against M. *leprae* antigens.

The skin lesions of leprosy are characterized by granuloma formation: a collection of lymphocytes and macrophages. The granulomatous lesions of tuberculoid leprosy are characterized by a predominance of CD4+ T cells, some of which are found surrounding macrophages, presumably interacting to restrict or eliminate the bacteria. In contrast, lepromatous lesions are characterized by a predominance of CD8+ lymphocytes, though in no discernible organization, and numerous macrophages loaded with bacteria. Thus, in order to restrict the growth of the M. *leprae*, a cooperation between CD4+ T cells and macrophages may be required.

CMI also appears to be critical to the outcome of the immune response against another mycobacteria, M. *tuberculosis*. Unlike most other forms of tuberculosis (TB), TB pleuritis usually resolves without chemotherapy,[2] suggesting that local CMI results in effective clearance of the bacilli. A selective concentration of M. *tuberculosis*-specific CD4+ T cells

Interleukin-10, edited by Jan E. deVries and René de Waal Malefyt.
© 1995 R.G. Landes Company.

in the pleural space is characteristic of TB pleuritis, much like the *M. leprae*-specific immunity present in the CD4+ T cells in the tuberculoid form of leprosy.

Cytokines play a vital role in regulating the interactions between T cells and macrophages, therefore we have recently focused our studies on determining how cytokines may mediate immune activation in tuberculoid leprosy and TB pleuritis lesions and immunosuppression in lepromatous leprosy lesions.

DISTRIBUTION OF TH1 AND TH2 CYTOKINES IN MYCOBACTERIAL DISEASE

We have examined the possibility that the paradoxical relationship between CMI and antibody production in the immune response to leprosy could be explained by diverse cytokine patterns in lesions. RT-PCR analysis of cytokine mRNA expression by lesions revealed distinct cytokine patterns. The Th1 or Type 1 cytokine pattern, typified by IFN-γ, IL-2, and lymphotoxin, predominated in tuberculoid lesions, whereas the Th2 or Type 2 pattern, IL-4, IL-5, and IL-10, was most prominent in lepromatous lesions.[3] Examination of T-cell clones derived from leprosy lesions revealed that the CD4+ T-cells in leprosy lesions produce IFN-γ and the CD8+ T cells in lepromatous lesions produce IL-4.[4]

Leprosy is not a static disease, but a dynamic condition in which immunologic changes can occur. In some cases, lepromatous patients can upgrade their clinical state towards the tuberculoid pole, referred to as a reversal reaction. When the pre- and post-reversal reaction lesions were compared in the same patients, a conversion from a Type 2 to a Type 1 phenotype occurred, with increases seen in IFN-γ and IL-2 expression and decreases in IL-4, IL-5 and IL-10 expression.[5]

Using a similar RT-PCR approach, we investigated cytokine expression in tuberculosis pleuritis, and in addition we measured cytokine levels using ELISA. Type 1 cytokines IFN-γ and IL-2 predominated in the pleural fluid when compared to PBMC

from the same donors, whereas IL-4 was lower in pleural fluid when compared to PBMC. IL-10 however, was also more prominent in the pleural fluid.[6]

The data from both leprosy and TB suggest that distinct cytokine patterns appear to have a role at the site of disease in response to mycobacterial infection. Type 1 cytokines predominate in lesions with effective CMI and Type 2 cytokines predominate in lesions of patients that are unresponsive to the pathogen, but mount humoral responses. The role of IL-10 in TB pleuritis is unclear; it may be responsible for limiting inflammation in response to *M. tuberculosis*.

CELLULAR SOURCE OF IL-10 IN MYCOBACTERIAL INFECTION AND THE ANTIGENS RESPONSIBLE FOR STIMULATING ITS PRODUCTION

IL-10 was originally described as a factor produced by Th2 cells that inhibits the response of Th1 cells.[7] To determine whether IL-10 was secreted by *M. leprae*-reactive T cells, T-cell clones from leprosy lesions were stimulated with anti-CD3 antibodies, and IL-10 release was determined.[8] Minimal IL-10 release was detected from the predominant T cells found in tuberculoid lesions, Th1-like CD4+ T cells (314 ± 104 pg/ml of IL-10, n = 9) or Th2-like CD8+ T-cells (57 ± 54 pg/ml, n = 4), those found most commonly in lepromatous lesions.[4] However, some *M. leprae*-reactive T-cell clones produced greater than 500pg/ml of IL-10.[9] In contrast, IL-10 was prominent in supernatants from CD4+ Th2-like cells (1052 ± 416, n=7) and CD8+ Th1-like (982 ± 368, n=6) cells; these populations were not *M. leprae*-reactive, but obtained from other sources.

To identify the cellular source of IL-10 released in response to *M. leprae*, PBMC were cultured in the presence or absence of *M. leprae* and IL-10 mRNA was measured by PCR. IL-10 mRNA was detectable at higher levels in *M. leprae*-stimulated cultures than in unstimulated cultures.

Subsequently, subpopulations of PBMC were immunomagnetically sorted after stimulation with *M. leprae*. Figure 9.1 shows that IL-10 mRNA was detected in CD14⁺ cells (monocytes) and adherent cells, but not in CD4⁺, CD8⁺ or B cells. Furthermore, *M. leprae* increased IL-10 mRNA levels in highly purified monocytes. Similarly, stimulation of a CD14⁺ fraction of pleural fluid mononuclear cells from TB pleuritis caused an increase in IL-10 mRNA expression.[6] These data suggest that monocytes/macrophages are the major source of IL-10 production in response mycobacterial infection, as has been shown for the in vitro response to LPS.[10]

Lipoarabinomannan (LAM), a mycobacterial cell wall heteropolysaccharide with a phosphatidylinositol moiety at its non-reducing end, has been implicated in the inhibition of T cell and macrophage activity.[11,12] We used LAM and chemically modified fractions of LAM to examine whether LAM could elicit IL-10 produc-

tion.[13] LAM, lipomannan, and phosphatidyl inositol mannosides stimulated IL-10 mRNA expression and IL-10 protein, whereas a deacylated form of LAM failed to induce IL-10. These results indicate that LAM stimulates IL-10 production and that the lipid portion of the molecule is required for this activity.

BIOLOGIC FUNCTION OF IL-10 IN MYCOBACTERIAL DISEASE

In order to further study the role of IL-10 in the immunopathogenesis of leprosy, we examined its effect on T cell and cytokine responses in vitro. IL-10 was initially described as an inhibitor of cytokine synthesis[7] based on its ability to inhibit T-cell responses. Therefore, we wished to ascertain whether the endogenous IL-10 produced in response to *M. leprae* was responsible for suppression of T-cell responses in leprosy. This was investigated using neutralizing antibodies to IL-10 for in vitro lymphocyte proliferation assays.[8] Anti-

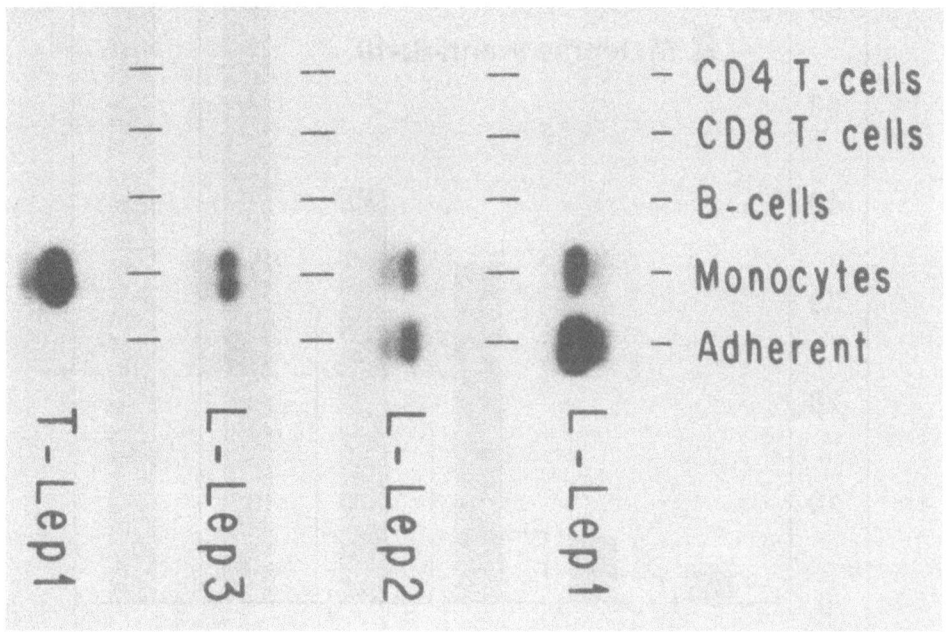

Fig. 9.1. M leprae-induced IL-10 mRNA from monocyte/macrophages. PBMC were cultured with M leprae for 14 h. Non adherent cells were isolated, immunomagnetically separated, and solubilized in guanidinium isothiocyanate. RNA was isolated, cDNA synthesized using reverse transcriptase, and PCR amplification was performed using IL-10-specific primers. PCR products were run on agarose gels, transferred to nylon membranes, and probed with a 32P-labeled IL-10-specific probe. T-Lep = tuberculoid leprosy; L-Lep=lepromatous leprosy.

IL-10 antibodies enhanced PBMC proliferation in both lepromatous and tuberculoid patients. Figure 9.2 shows representative proliferation assays from three lepromatous and three tuberculoid patients. Mean percentage increases in the presence of anti-IL-10: lepromatous patients = 85% (n=24), range=0 to 360%; tuberculoid patients=63% (n=18), range=0 to 410%. Although anti-IL-10 antibodies enhanced proliferation in normally unresponsive lepromatous patient PBMCs, in most cases the levels did not reach those of tuberculoid patients, suggesting that additional factors are responsible for the unresponsiveness in lepromatous patients.

TB patients with concomitant HIV infection (HIV+) have decreased T-cell proliferative responses to *M. tuberculosis* when compared to HIV-TB patients (mean Δ cpm HIV+18,895 ± 5,361, 62,593 ± 15,626, p < 0.01). We examined the possibility that this decrease in T-cell responses was due to the release of IL-10, given that we have previously shown that mycobacterial LAM stimulates IL-10 release.[13] Figure 9.3 shows that neutralizing antibodies to IL-10 augmented T-cell proliferation to *M. tuberculosis* by 60% or greater in eight of ten HIV+TB patients (mean percent increase=130, range 0 to 520%), although the responses did not increase to the level

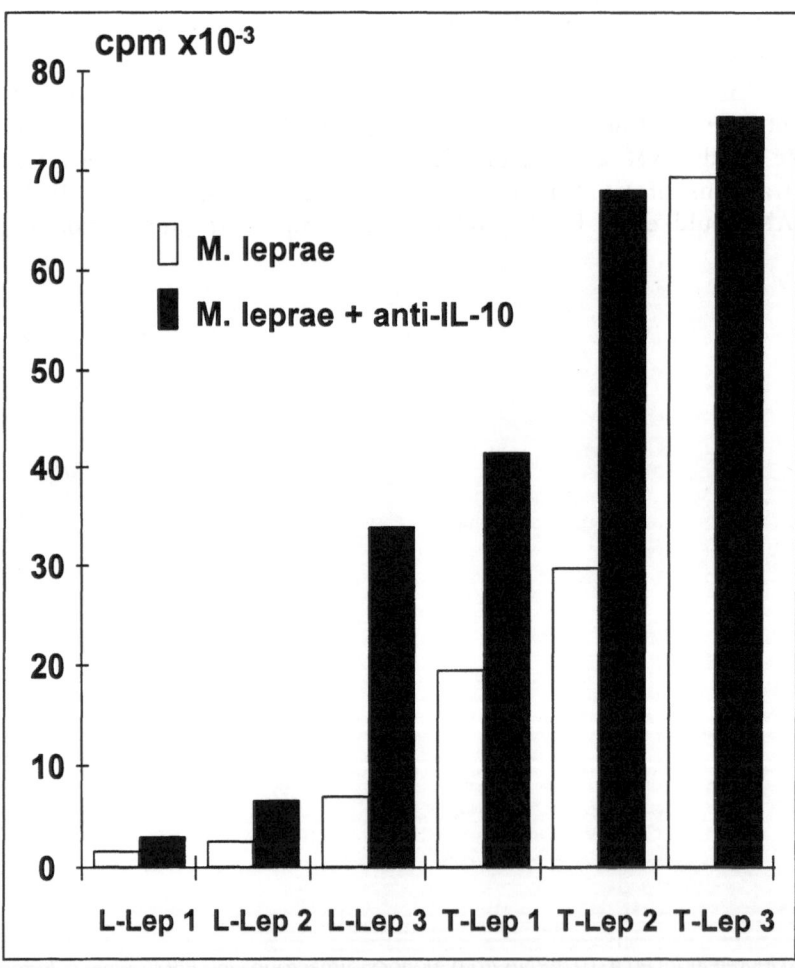

Fig. 9.2. M. leprae-induced IL-10 release inhibits lymphocyte proliferation. PBMC were cultured with M. leprae and neutralizing anti-IL-10 antibodies. Five days after initial culture, cells were pulsed with ³H-thymidine, harvested 18 to 24 h later, and the ³H-thymidine incorporation was measured. Values are reported as the mean cpm in the presence or absence of anti-IL-10. Isotype-matched, control antibodies had no effect on PBMC proliferation to M. leprae. T-Lep = tuberculoid leprosy; L-Lep = lepromatous leprosy.

of HIV⁻ TB patients. These data indicate that endogenous IL-10 release is responsible in part for the weak T-cell responses to *M. tuberculosis* in HIV⁺ TB patients. Together, these results suggest that the mycobacteria suppress T-cell proliferative responses through the release of IL-10.

M. LEPRAE-INDUCED IL-10 SECRETION SUPPRESSES CYTOKINE RELEASE

In addition to the effects of IL-10 on T-cell proliferation, we examined the effect of endogenous IL-10 production on *M. leprae* induction of cytokine release.[8] IFN-γ facilitates macrophage killing of mycobacteria[14,15] and stimulates macrophages to express HLA-DR and ICAM-1, which facilitate T-cell responses.[16] In five out of six donors, *M. leprae*-induced IFN-γ release from

PBMC was increased after treatment with anti-IL-10 indicating that endogenous IL-10 production inhibited IFN-γ release (Fig. 9.4).

Two other important cytokines in the immune response to mycobacteria are TNF-α and GM-CSF, which contribute to granuloma formation and killing of mycobacteria. These cytokines are predominantly produced by macrophages in response to mycobacteria in vitro.[13] TNF-α (mean=1074 pg/ml) and GM-CSF (mean=34 pg/ml) were released after *M. leprae* stimulation of PBMC, whereas TNF-α and GM-CSF release were undetectable in unstimulated cultures. Addition of anti-IL-10 antibodies to cultures increased *M. leprae*-induced TNF-α and GM-CSF release in 15 of 17 donors.[8] In summary, production of IL-10 in response to *M. leprae* may result in inhibition of anti-mycobacterial cytokine synthesis.

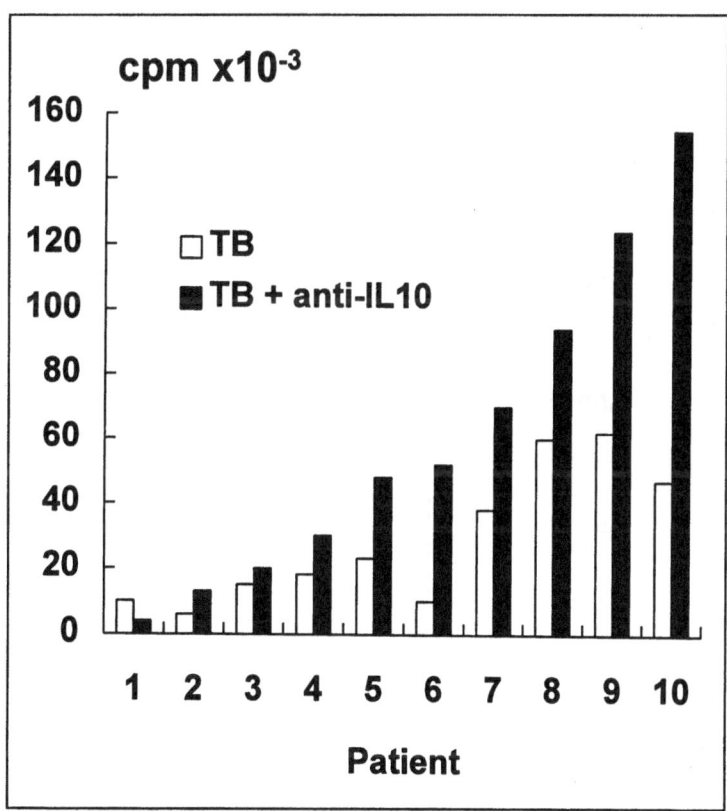

Fig. 9.3. Effect of neutralizing antibodies to IL-10 on the M. tuberculosis-induced proliferative response of PBMC from HIV-infected tuberculosis patients. Proliferative responses to M. tuberculosis were determined in the absence or presence of neutralizing antibodies to IL-10. Values are reported as the mean cpm of triplicate cultures.

IL-10 INHIBITS IL-12 PRODUCTION

Of fundamental immunologic importance are the factors which influence the nature of the cytokine response. One important factor is IL-12, which selectively induces the Th1 cytokine pattern.[17-21] IL-12 is primarily produced by activated monocytes[22] as a heterodimeric protein composed of p35 and p40 subunits[23] and acts as a T-cell growth factor.[17] Our recent studies indicate that IL-12 p40 is more prominently expressed in tuberculoid as compared to lepromatous leprosy lesions.[24] One possible explanation

for the low levels of IL-12 p40 mRNA in lepromatous lesions could be the presence of high levels of Type 2 cytokines.[3]

To investigate this possibility, we measured IL-12 release from *M. leprae*-stimulated monocytes by antibody capture bioassay. IL-12 was undetectable in cultures grown in the absence of *M. leprae*. However, *M. leprae*-stimulated IL-12 release in tuberculoid (21.7 ± 11.2 pg/ml, n=5) and lepromatous (20.0 ± 3.9, n=5) patients in equivalent amounts, indicating that the monocytes of lepromatous patients displayed no inherent defect in their ability

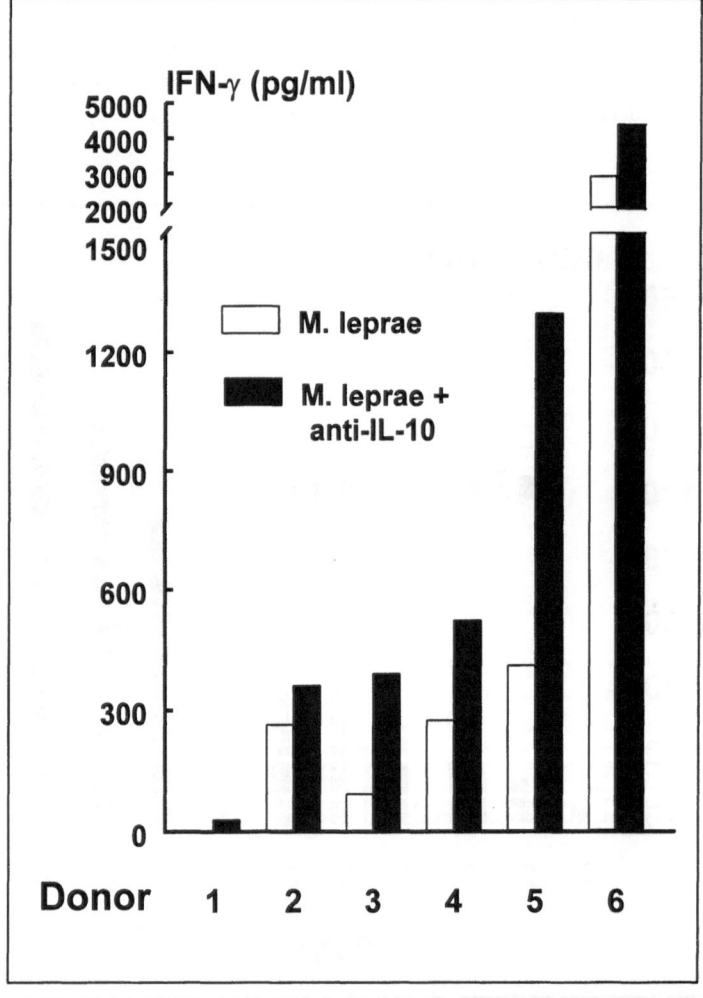

Fig. 9.4. M. leprae-induced IL-10 secretion suppresses IFN-γ-release. PBMC were cultured with M. leprae in the presence or absence of neutralizing anti-IL-10 antibodies. IFN-γ was measured in supernatants (24 h) by ELISA using cytokine-specific antibodies. Values (n = 6) are reported as pg/ml.

to produce IL-12. To resolve the paradox between in vivo and in vitro IL-12 production, we examined the effects of IL-10 on IL-12 production. rIL-10 inhibited *M. leprae*-specific IL-12 release to below detectable levels indicating that IL-10, produced in lepromatous lesions, may suppress the local production of IL-12 (Fig. 9.5). The downregulatory effects of IL-10 on IL-12 production have also been reported in separate systems.[18,25]

In addition, neutralizing IL-10 antibodies enhanced *M. leprae*-induced PBMC release of IL-12 in five of six leprosy patients. Furthermore, anti-IL-10 enhanced *M. tuberculosis*-induced PBMC release of four of four donors indicating that endogenous IL-10 inhibits the production of IL-12 (unpublished observations). The ability of IL-10 to down regulate IL-12 production may lead to inhibition of Th1 cytokine responses.

IL-10 INHIBITS CD1 EXPRESSION AND SUBSEQUENT *M. LEPRAE*-SPECIFIC PROLIFERATION OF αβ T-CELL RECEPTOR CD4, CD8 DOUBLE NEGATIVE T-CELLS

There is a small population of peripheral T-cells which express the αβ T-cell receptor (TCR), but lacks both CD4 and CD8 (CD4⁻, CD8⁻, double negative (DN) αβTCR). Double negative T cells have been identified in the skin[26] and peripheral blood

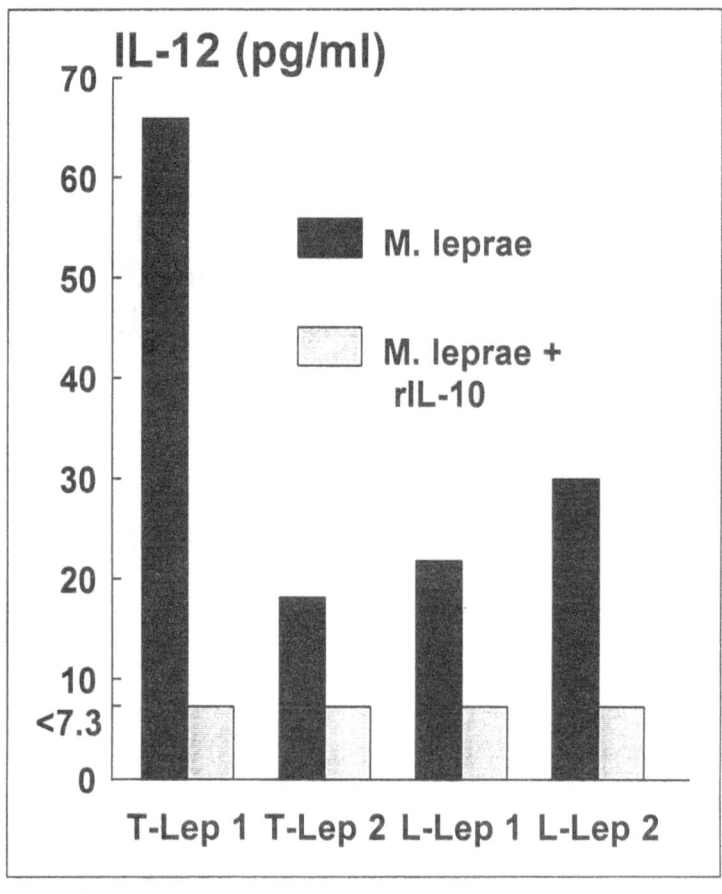

Fig. 9.5. Inhibition of M. leprae-stimulated IL-12 release in vitro by rIL-10. Supernatants from adherent PBMC stimulated with M. leprae (24 h) in the presence or absence of rIL-10 (100 U/ml) were examined for IL-12 release by bioassay. Values are expressed as IL-12 heterodimer in pg/ml. T-Lep = tuberculoid leprosy; L-Lep = lepromatous leprosy.

of normal individuals[27] and at increased levels in some patients with autoimmune diseases.[28,29] DNαβ T cells responsive to *M. tuberculosis* and *Escherichia coli* have been isolated from the peripheral blood of normal donors, suggesting a role for these cells in combating infection.[30,31]

Antigen-responsive DNαβ T cells have been shown to be restricted by a non-polymorphic MHC class I-like molecule, CD1.[30] Four forms of human CD1 have been identified, CD1a, CD1b, and CD1c being more similar to each other in their amino acid sequences than CD1d.[32] Consistent with a potential role for DNαβ T cells in cell-mediated immunity (CMI) to infection, CD1 molecules are located at particular anatomic sites. For example, CD1 is expressed on Langerhans cells and other an-

tigen presenting cells (APC) in skin.[33] Investigation of the DNαβ T-cell-CD1 interaction has been facilitated by the demonstration that CD1 can be induced on monocytes by stimulation with GM-CSF.[34] We investigated the role of CD1 and DNαβ T-cells in the immune response to leprosy.

Although CD1 was weakly expressed in the skin lesions of lepromatous patients, it could be induced on PBMC from those patients by in vitro culture with either GM-GSF alone or GM-CSF plus IL-4. CD1a, CD1b, and CD1c expression was equally inducible on monocytes of tuberculoid and lepromatous patients. We also found that similar levels of CD1 expression could be induced on PBMC of normal donors.

We considered that the presence of Type 2 cytokines in lepromatous lesions

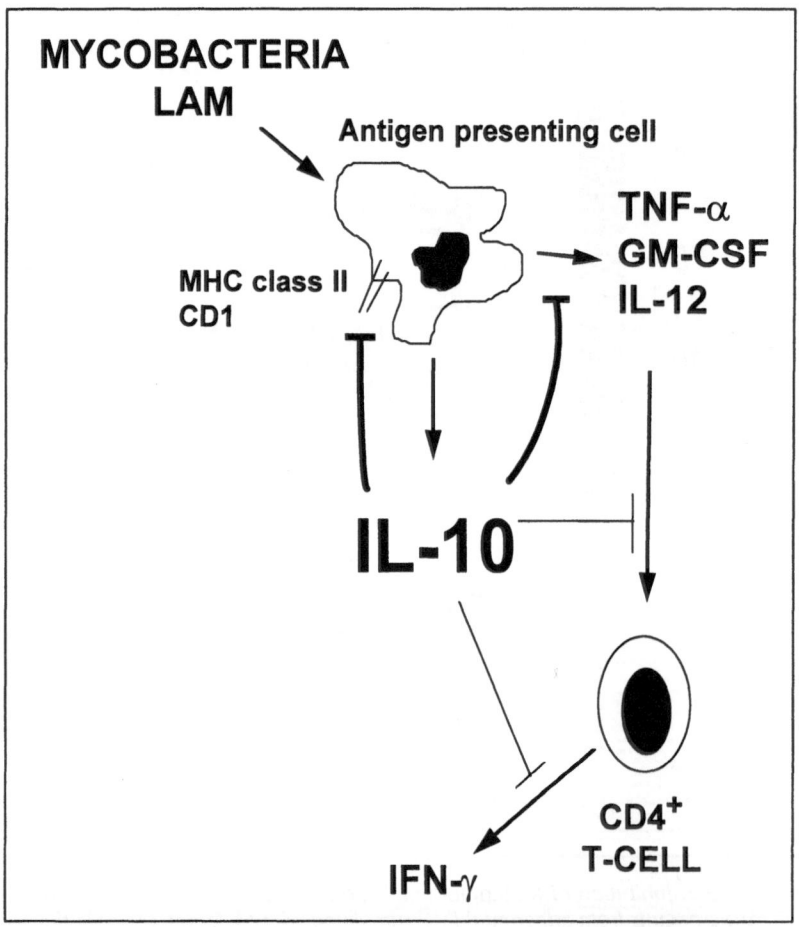

Fig. 9.6. Model for the immunosuppressive role of IL-10 in mycobacterial infection. Thick lines indicate direct effects, thin lines indicate indirect effects of IL-10.

could result in the downregulation of CD1 expression. IL-10 is known to inhibit the expression of MHC class II[35] and IL-10 mRNA is present in higher levels in lepromatous lesions as compared to tuberculoid lesions.[3] Therefore, rIL-10 was added to the in vitro culture system to determine whether it could inhibit CD1 expression. IL-10 almost completely blocked GM-CSF plus IL-4-induced CD1 expression.

We also demonstrated that IL-10 inhibition of CD1 expression on monocytes inhibited antigen-presentation to DNαβ T cells. PBMC were cultured with media alone or with cytokines, and the treated cells were then used to present antigen to DNαβ T cells. rIL-10 treatment of APC inhibited the antigen-presenting capacity of the monocytes to DNαβ T cells by more than 80%. Therefore, IL-10 inhibition of CD1 expression precludes antigen presentation to DNαβ T cells.

CONCLUDING REMARKS

Effective CMI to mycobacterial infection seems to require IFN-γ-producing T cells interacting with macrophages. Cytokines play an important role in promoting or inhibiting these interactions, thereby leading to elimination or dissemination of the pathogen. We studied the role of the immunosuppressive cytokine, IL-10 in mediating immunosuppression in mycobacterial disease. We found that: (1) IL-10 was prominent in lesions of immunologically unresponsive lepromatous leprosy patients; (2) monocytes produced IL-10 in response to mycobacteria; (3) LAM stimulated IL-10 release and the lipid moiety of LAM was required for IL-10 release; (4) IL-10 inhibited T-cell proliferation and T-cell and macrophage cytokine production in response to mycobacteria; and (5) IL-10 inhibited CD1 expression and subsequent antigen presentation to CD4-, CD8- DNαβ T cells.

The data from these studies suggest that on the whole, IL-10 upsets the cooperation between T cells and macrophages in CMI. A model for the role of IL-10 in leprosy is shown in Figure 9.6. LAM-induced IL-10 release inhibits the macrophage from producing TNF-α and GM-CSF cytokines, which activate macrophages to phagocytize bacteria. IL-10 inhibits MHC class II and CD1 expression on antigen presenting cells, thereby inhibiting activation of IFN-γ-producing CD4+ and DNαβ T cells at the site of infection. IFN-γ-producing T cells are also inhibited by IL-10 through down regulation of IL-12 production. It will be interesting to know the kinetics of IL-10 production in response to mycobacterial infection. Since macrophages release IL-10 in a non-specific manner, release of IL-10 may be an early event in the response to mycobacteria and may therefore play a pivotal role in determining the level of CMI to mycobacteria and the eventual outcome of infection.

REFERENCES

1. Ridley DS, Jopling WH. Classification of leprosy according to immunity. A five-group system. Int J Lepr 1966; 34:255-273.
2. Roper WH, Waring JJ. Primary serofibrinous pleural effusion in military personnel. Amer Rev Tuberc 1955; 71:616-634.
3. Yamamura M, Uyemura K, Deans RJ et al. Defining protective responses to pathogens: cytokine profiles in leprosy lesions. Science 1991; 254:277-279.
4. Salgame P, Abrams JS, Clayberger C et al. Differing lymphokine profiles of functional subsets of human CD4 and CD8 T-cell clones. Science 1991; 254:279-282.
5. Yamamura M, Wang X-H, Ohmen JD et al. Cytokine patterns of immunologically mediated tissue damage. J Immunol 1992; 149:1470-1475.
6. Barnes PF, Lu S, Abrams JS, et al. Cytokine production at the site of disease of human tuberculosis. Infect Immun 1993; 61:3482-3489.
7. Fiorentino DF, Bond MW, Mosmann TR. Two types of mouse T-helper cell. IV. Th2 clones secrete a factor that inhibits cytokine production by Th1 clones. J Exp Med 1989; 170:2081-2095.
8. Sieling PA, Abrams JS, Yamamura M et al. Immunosuppressive roles for interleukin-10

and interleukin-4 in human infection: In vitro modulation of T cell responses in leprosy. J Immunol 1993; 150:5501-5510.

9. Mutis T, Kraakman EM, Cornelisse YE et al. Analysis of cytokine production by Mycobacterium-reactive T cells. Failure to explain Mycobacterium leprae-specific nonresponsiveness of peripheral blood T cells from lepromatous leprosy patients. J Immunol 1993; 150:4641-4651.

10. de Waal Malefyt R, Abrams J, Bennett B et al. Interleukin 10 (IL-10) inhibits cytokine synthesis by human monocytes: An autoregulatory role of IL-10 produced by monocytes. J Exp Med 1991; 174: 1209-1220.

11. Moreno C, Mehlert A, Lamb J. The inhibitory effects of mycobacterial lipoarabinomannan and polysaccharides upon polyclonal and monoclonal human T cell proliferation. Clin Exp Immunol 1988; 74:206-210.

12. Sibley LD, Hunter SW, Brennan PJ et al. Mycobacterial lipoarabinomannan inhibits gamma interferon-mediated activation of macrophages. Infect Immun 1988; 56: 1232-1236.

13. Barnes PF, Chatterjee D, Abrams JS et al. Cytokine production induced by Mycobacterium tuberculosis lipoarabinomannan. Relationship to chemical structure. J Immunol 1992; 149:541-547.

14. Rook GAW, Steele J, Fraher L et al. Vitamin D3, gamma interferon, and control of proliferation of Mycobacterium tuberculosis by human monocytes. Immunology 1986; 57:159-163.

15. Nathan CF, Kaplan G, Levis WR et al. Local and systemic effects of intradermal recombinant interferon-gamma in patients with lepromatous leprosy. N Engl J Med 1986; 315:6-15.

16. Dustin ML, Singer KH, Tuck DT et al. Adhesion of T lymphoblasts to epidermal keratinocytes is regulated by interferon-gamma and is mediated by intercellular adhesion molecule 1 (ICAM-1). J Exp Med 1988; 167:1323-1340.

17. Gately MK, Desai BB, Wolitsky AG et al. Regulation of human lymphocyte proliferation by a heterodimeric cytokine, IL-12 (cytotoxic lymphocyte maturation factor). J Immunol 1991; 147:874-882.

18. Hsieh C, Macatonia SE, Tripp CS et al. Development of Th1 CD4+ T cells through IL-12 produced by Listeria-induced macrophages. Science 1993; 260:547-549.

19. Heinzel FP, Schoenhaut DS, Rerko RM et al. Recombinant interleukin 12 cures mice infected with Leishmania major. J Exp Med 1993; 177:1505-1509.

20. Sypek JP, Chung CL, Mayor SEH et al. Resolution of cutaneous leishmaniasis: interleukin 12 initiates a protective T helper type 1 response. J Exp Med 1993; 177: 1797-1802.

21. Seder RA, Gazzinelli R, Sher A et al. Interleukin 12 acts directly on CD4+ T cells to enhance priming for interferon-gamma production and diminishes interleukin 4 inhibition of such priming. Proc Natl Acad Sci USA 1993; 90:10188-10192.

22. D'Andrea A, Rengaraju M, Valiante NM et al. Production of natural killer cell stimulatory cell factor (interleukin 12) by peripheral blood mononuclear cells. J Exp Med 1992; 176:1387-1398.

23. Stern AS, Podlaski FJ, Hulmes JD et al. Purification to homogeneity and partial characterization of cytotoxic lymphocyte maturation factor from human B-lymphoblastoid cells. Proc Natl Acad Sci USA 1990; 87:6808-6812.

24. Sieling PA, Wang X-H, Gately MK et al. IL-12 regulates T helper Type 1 cytokine responses in human infectious disease. 1993, J Immunol. in press..

25. D'Andrea A, Aste-Amezaga M, Valiante NM et al. Interleukin 10 (IL-10) inhibits human lymphocyte interferon gamma-production by suppressing natural killer cell stimulatory factor/IL-12 synthesis in accessory cells. J Exp Med 1993; 178:1041-1048.

26. Groh V, Fabbi M, Hochstenbach F et al. Double-negative (CD4-CD8-) lymphocytes bearing T-cell receptor alpha and beta chains in normal human skin. Proc Natl Acad Sci USA 1989; 86:5059-5063.

27. Lanier LL, Weiss A. Presence of Ti (WT31) negative T lymphocytes in normal blood and thymus. Nature 1989; 324:268-270.

28. Shivakumar S, Tsokos GC, Datta SK. T-cell receptor alpha/beta expressing double-nega-

tive (CD4-/CD8-) and CD4⁺ T helper cells in humans augment the production of pathogenic anti-DNA autoantibodies associated with lupus nephritis. J Immunol 1989; 143:103-112.

29. Brooks EG, Wirt DP, Goldblum RM et al. Double negative (CD4- CD8⁻) T cells with an α/β T cell receptor. Non-MHC-restricted cytolytic activity and lymphokine production. J Immunol 1990; 144:4507-4512.

30. Porcelli S, Morita CT, Brenner MB. CD1b restricts the response of human CD4-8- T lymphocytes to a microbial antigen. Nature 1992; 360:593-597.

31. Dellabona P, Casorati G, Friedli B et al. In vivo persistence of expanded clones specific for bacterial antigens within the human T cell receptor alpha/beta CD4-8- subset. J Exp Med 1993; 177:1763-1771.

32. Calabi F, Bradbury A. The CD1 system. Tissue Antigens 1991; 37:1-9.

33. Meunier L, Gonzalez-Ramos A, Cooper KD. Heterogeneous populations of class II MHC⁺ cells in human dermal cell suspensions. Identification of a small subset responsible for potent dermal antigen-presenting cell activity with features analogous to Langerhans cells. J Immunol 1993; 151: 4067-4080.

34. Kasinrerk W, Baumruker T, Majdic O et al. CD1 molecule expression on human monocytes induced by granulocyte-macrophage colony-stimulating factor. J Immunol 1993; 150:579-584.

35. de Waal-Malefyt R, Haanen J, Spits H et al. Interleukin 10 (IL-10) and viral IL-10 strongly reduce antigen-specific human T-cell proliferation by diminishing the antigen-presenting capacity of monocytes via down-regulation of class II major histocompatibilty complex expression. J Exp Med 1991; 174:915-924.

CHAPTER 10

IL-10 IN HUMAN LEISHMANIASIS

Edgar M. Carvalho

INTRODUCTION

Leishmaniasis is endemic in many tropical and subtropical countries. It is estimated that worldwide 400 million individuals are exposed to leishmania infection, with an annual incidence of 600,000 and a prevalence of 12 million.[1] The disease is caused by protozoa of the genus leishmania (kinetoplastida: tripanosomatidae) that are transmited to humans by the bite of infected phlebotomine sand flies. With the exception of India, where man is the main target for leishmania infection, leishmaniasis is a zoonotic disease and humans are only incidental hosts in the parasite's life cycle. The clinical manifestations of leishmaniasis are variable and are related, in part, to the strain of the infecting agent, the environment and the host immunological response. Four different clinical forms of leishmaniasis are well characterized: cutaneous leishmaniasis, mucosal leishmaniasis, diffuse cutaneous leishmaniasis and visceral leishmaniasis.

There are two distinct stages in the life cycle of leishmania: a motile flagellated promastigote stage that lives extracellularly within the alimentary tract of the sandfly vector, and a non-motile amastigote stage that lives within macrophages of the mammalian host.[2] In the sandfly and in culture in the stationary phase of growth, biological, chemical and antigenic changes occur which are associated with changes in infectivity.[3] After leishmania inoculation via the human skin, the promastigote form penetrates into macrophages. Although fresh human sera is capable of destroying the promastigote form of leishmania,[4] antibody and complement play no role as a defense mechanism against leishmania.[5,6] The infecting promastigote form (metacyclic promastigote) is resistant to fresh sera[7-9] and complement is not capable of destroying the amastigote form of the parasite. Promastigote and amastigote forms of leishmania may be killed by neutrophils and peripheral blood monocytes. The

Interleukin-10, edited by Jan E. deVries and René de Waal Malefyt.

killing is related to the toxic effects of oxigen metabolites produced during the macrophage induced respiratory burst[10,11] and to the non-oxidative killing effect of lysosomal hydrolases, or nitric oxide production from L.-arginine.[12] In contrast to neutrophils and monocytes, macrophages have decreased ability to produce such components and, are therefore, more susceptible to leishmania infection. Additionally, there are various mechanisms developed by leishmanial parasites to evade host defenses, allowing these organisms to survive in the immunologically competent host.[13-15] Furthermore, salivary gland material from *Lutzomya longipalpis*, enhance leishmania infectivity both through an inhibitory effect on the macrophages' ability to present parasite antigens to specific T cells, and by downregulating the ability of those cells to produce hydrogen peroxide in response to an activating stimulus from IFN-γ.[16,17]

Although macrophages may have the innate ability to kill leishmania, studies in experimental animals and in vitro studies with human cells have pointed out the need for a T-cell response to control leishmania infection.[5,6,18] Protective immunity against leishmania has been related, predominantly, to IFN-γ producing CD4+ Th1 helper T cells. IFN-γ is known as the macrophage activating factor and addition of this cytokine to leishmania infected macrophages results in leishmania killing.[10] There are also indications of the participation of cytotoxic T cells in the control of leishmania infection. Extremely susceptible BALB/c mice have 80 to 90% of their antigen specific T cells of the L3T4+ phenotype, whereas resistant CBA mice have antigen-specific cells divided between L3T4 and LyT2+ cells.[19] Additionally, anti LyT2+ treatment blocks the induction of resistance in intravenously immunized BALB/c mice[20] and CD8+ T cells were able to mediate protective immunity in murine leishmaniasis following deletion of CD4+ T cells in BALB/c animals.[21] Furthermore, the healing of cutaneous leishmaniasis ulcers is associated with an increased number of leishmania specific CD8+ T cells.[22] Cytotoxic response may lyse infected target cells thereby leading either to the killing of leishmania or to the liberation of the leishmania. In the latter case, the leishmania may proceed to infect activated cells that are in turn capable of killing the parasite.

In both mechanisms of leishmania killing, cytokines produced by monocytes and T cells play fundamental roles. These observations were initially well documented in experimental studies with both leishmania susceptible and resistant mice. Host resistance to leishmania infection is associated with activation and differentiation of effective CD4+ helper T (Th) cells, Th1 subset, whose cytokine secretion pattern is formed by IL-2, IFN-γ and lymphotoxin.[23-26] In contrast, progressive and fatal infection is associated with Th2 CD4+ cell response and production of IL-4, IL-5, IL-6 and IL-10.[24-27] The mechanisms leading to the preferential induction of a distinct Th cell subset are still not completely known. Factors which may play a role in modulating the cell response during leishmania infection include:

1. The type of antigen presenting cell;
2. The type of the antigen that is recognized by the T cell;
3. The cytokine environment during the initial events of the immune response;
4. The different signalling mechanisms used by the cell subsets after interaction with the antigen.

Macrophages act as both host cells for leishmania and as antigen presenting cells and also produce cytokines that modulate, in association with T cells, immunological response. They therefore play key roles in determining whether leishmania infection will be controlled, or will progress to disease. For instance, IL-1 and TNF-α are cytokines produced by macrophages and necessary for the immunological response to, and control of, leishmania infection. On the other hand, IFN-γ produced by T cells activates macrophages and induces leishmania killing. In contrast, production of TGF-β and IL-10 may help the parasite to survive and multiply within macrophages.[28,29]

Leishmania infection induces the production of active TGF-β. Animals resistant to *L.amazonensis* infection, such as the C57BL/6, or to *L.braziliensis*, such as the BALB/c mice, show disease progression when TGF-β is administered.[30] In contrast, neutralizing monoclonal antibody to TGF-β has a protective effect in the course of *L.amazonensis* infection in the BALB/c mice.[30] The roles of IL-10 in the regulation of the immunological responses in leishmania and other infectious diseases such as schistosomiasis,[31] Chagas' disease,[32] leprosy[33] and AIDS[34] have recently been studied in experimental animals and humans. IL-10 is produced by macrophages, Th2 cells, B cells and mast cells[35,36] and have a predominantly down regulatory effect in the immune response. IL-10 may affect antigen presentation by its action on dendritic cells and macrophages[37,38] may deactivate macrophages preventing killing of intracellular pathogens[39,40] and downregulating T-cell responses.[41] The present chapter deals with the role of IL-10 in modulating the immune response in the different clinical forms of leishmaniasis: visceral leishmaniasis, diffuse cutaneous leishmaniasis and tegumentary (cutaneous and mucosal) leishmaniasis.

IL-10 IN VISCERAL LEISHMANIASIS

Visceral leishmaniasis, or Kala-azar, is the systemic form of leishmania infection. The disease is caused by leishmania of the complex *L.donovani* (*L.donovani, L.chagasi* and *L.infantum*). Fever, weight loss, hepatosplenomegaly, anemia, leukopenia and hyperglobulinemia are the main clinical features of this disease. Immunologically, visceral leishmaniasis is characterized by an absence of T-cell response to leishmania antigen[42-45] and polyclonal B-cell activation.[46,47]

In endemic areas of visceral leishmaniasis, infected subjects may, or may not, develop classic signs and symptoms of visceral leishmaniasis.[48] Capacity to produce IL-2 and IFN-γ is associated with asymptomatic or subclinical self-healing visceral leishmaniasis.[18] In contrast, individuals whose lymphocytes do not proliferate and, therefore, do not produce IFN-γ when stimulated by leishmania antigen, will develop acute visceral leishmaniasis or a subclinical infection that progresses to classical visceral leishmaniasis.[18] Immunological abnormalities can be found in monocyte and T-cell function, such as diminished production of TNF-α and IL-1 after LPS or lysteria stimulation,[49] decreased production of IL-2 and IFN-γ by lymphocytes under leishmania antigen stimulation,[50] absence of delayed type hypersensitivity to leishmania antigen[51,52] and decreased capability of T cells to activate macrophages and kill leishmania.[53] All these abnormalities may account for parasite multiplication and progression of the disease. The possibility of the occurence of Th2 activation in visceral leishmaniasis has been suported by the documentation of high titers of antibody against leishmania antigen, the predominance of IL-4 over IFN-γ levels in visceral leishmaniasis sera,[54] isolation of Th2 clones from cured visceral leishmaniasis patients,[55] and evidence of mRNA for IL-4 and IL-10 in mononuclear cells from lymph nodes[56] spleen[57] and peripheral blood.[29] In fact, several abnormalities observed in visceral leishmaniasis may be associated with IL-10 production. IL-10 inhibits cytokine production by macrophages[58,59] and decreases the capability of antigen presenting to T cells by reducting, the expression of MHC class II.[37] These abnormalities have been documented in macrophages infected by *L.donovani*[60] and in monocytes from visceral leishmaniasis patients.[49] Additionally IL-10 is capable of suppressing lymphocyte proliferation,[41,61] IFN-γ production[62] and delayed type hypersensitivity.[63] The strongest evidence that IL-10 is involved in down regulate immunological responses in visceral leishmaniasis is the documentation that neutralizing human monoclonal antibody, anti IL-10, restores lymphocyte proliferation and IFN-γ production in visceral leishmaniasis (Figure 10.1). While the production of IFN-γ in supernatant of cultures stimulated with leishmania antigen is absent, in cultures stimulated by leishmania antigen plus anti IL-10 significative IFN-γ production is achieved. In terms of lymphocyte proliferative response, anti-IL-10 acts synergistically with anti-IL-4 to enhance lymphocyte blastogenesis.[29] The role of IL-10 in downregulating T-cell response in visceral leishmaniasis was further documented by the ability of IL-10 to suppress lymphocyte proliferation and IFN-γ production in subjects cured of visceral leishmaniasis. IL-10 levels are quite variable in supernatant of mononuclear cell cultures from visceral leishmaniasis patients stimulated with leishmania antigen. In several cases, although IL-10 production is strong in unstimulated cultures, IL-10 levels are undetectable in cultures stimulated with parasite antigen. This variation may be related to the auto regulatory capability of IL-10.[59]

Produced in the initial phase of the immune response, IL-10 may inhibit production and action of cytokines capable to drive a Th1 response such as IL-12. IL-12 is predominantly produced by macrophages and acts on NK cells and T cells that are cells involved in IFN-γ production and activation of macrophages to kill intracellular pathogens. Production of IL-12 by macrophages and consequently IFN-γ production by NK cells are key components for the host to have a Th1 type response.[64]

Experimental studies in SCID mice infected with intracellular pathogens such as leishmania or *T. gondii* showed that intracellular pathogens induce macrophages to produce IL-12 and TNF-α and that such cytokines induce splenocytes to produce IFN-γ, TNF-α and IL-12.[65] In this case production of IFN-γ is abrogated by antibodies to IL-12. When macrophages phagocytose or are actively invaded by pathogens, IL-12 is induced relatively rapidly, with a peak around 4-8hs and induces the generation of IFN-γ by NK cells.[66] IFN-γ then activates macrophage production of hydrogen peroxidase and nitrate synthase that enables macrophages to restrict multiplication of intracellular pathogens. IL-12 and latter IFN-γ production will then induce a Th1 activation and consequently more IFN-γ production by Th1 cell. IL-10 is also produceed by macrophage and down regulates IFN-γ production by affecting macrophages (inhibition of TNF-α and IL-1 production), NK cells by interfering with the stimulatory effect of IL-12 and TNF-α and Th1 cells by suppress cytokine production.

In visceral leishmaniasis IL-12 has similar effects as the monoclonal antibodies to IL-10 in the context of restoring lymphocyte proliferation and IFN-γ production by mononuclear cells from patients

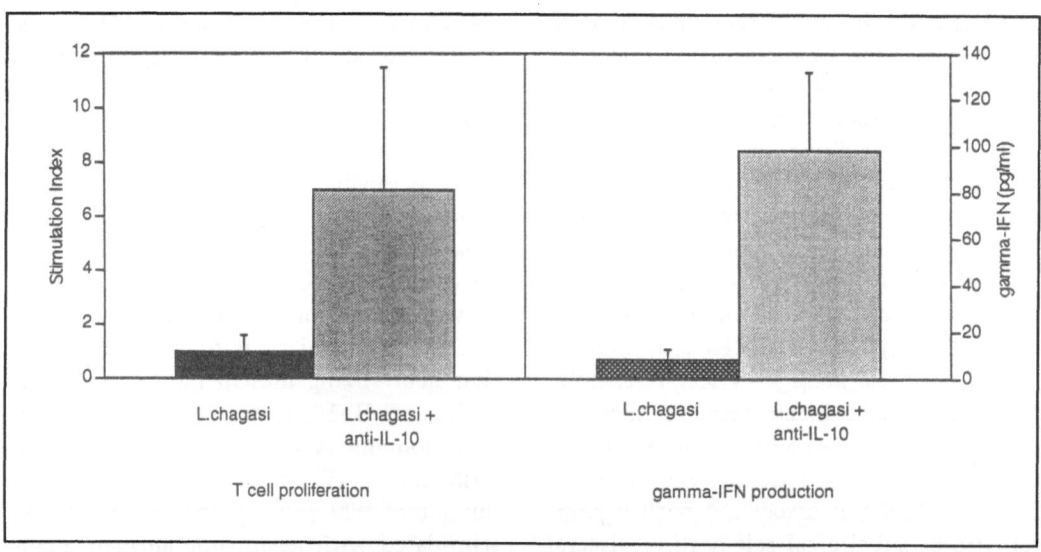

Fig. 10.1 *Anti-IL-10 restores lymphocyte proliferation and INF-γ production in visceral leishmaniasis.*

with visceral leishmaniasis. Therefore the imbalance of these antagonistic cytokines may be involved in the pathogenesis of visceral leishmaniasis. For instance IL-10 abrogates the IFN-γ and IL-2 production mediated by IL-12 and monoclonal antibody to IL-12 abrogates restoration of T-cell response modulated by anti-IL-10.

As the initial events after the contact of the infecting agent with host cells are critical in determining the type of the immune response, and since IL-10 and IL-12 are produced by antigen presenting cells, it is conceivable that such cytokines are fundamental to indicate the clinical course of leishmania infection. In such case, there would be a prodominant production of IL-10 after interaction of the parasite with macrophages and consequently a Th2 type response in subjects which are susceptible to develop visceral leishmaniasis. The main result of it would be a decrease in IFN-γ production and in the cytotoxic response, mechanisms which are involved in leishmania killing.

IL-10 IN DIFFUSE CUTANEOUS LEISHMANIASIS

The clinical forms of cutaneous leishmaniasis depend on both host immunological responses and parasite virulence. They range from a hyper-responsive status observed in mucosal leishmaniasis to unresponsive diffuse cutaneous leishmaniasis. Diffuse cutaneous leishmaniasis (DCL) is clinically characterized by the presence of multiple nodules or plaques without ulceration consisting of abundant vacuolated macrophages full of leishmania amastigotes.[67,68] There are only a few lymphocytes in the lesion and immunologically T cells from DCL patients do not proliferate or produce IFN-γ when stimulated by leishmania antigen. It is generally admitted that DCL results from a lack of cell mediated immunity to parasite antigens leading to uncontrolled parasite growth.[69] In contrast to the depression of T-cell responses, specific antibody production is preserved and high titers of antibody to leishmania antigen are found in this disease. Responses to

antimonial therapy, or to others drugs used in the treatment of leishmaniasis, such as amphotericin B, penthamidine and aminosidine, is partial. Disappearance of the lesions is not complete and relapse is always observed.

Immunological abnormalities found in diffuse cutaneous leishmaniasis are more similar to those observed in visceral leishmaniasis. However, the restoration of T-cell response after therapy, a finding observed early after treatment for visceral leishmaniasis, usually does not occur with DCL. In a few patients, evidence of lymphocyte blastogenesis following therapy has been documented. This phenomena is, however, transitory and sometimes only minimal enhancement of lymphocyte proliferation is observed.[70] In the majority of cases, the delayed type hypersensitivity and IFN-γ production remain absent even in patients that exhibited clinical remission post therapy. The abnormalities in cell mediated immune response are antigen specific and lymphocyte proliferation and IFN-γ production can be documented in cultures stimulated by unrelated antigens and mitogens. Early studies evaluating the mechanism of depressed T-cell responses in DCL have pointed out the role of macrophages in this phenomena. They showed enhancement of lymphocyte proliferation in culture depleted of macrophages and evidence of cellular responses when T cells from DCL patients were co-cultured with macrophages of MHC compatible relatives.[71] Immunological studies performed with cells obtained from DCL lesions have evaluated the cytokine profile at the molecular level. They have shown evidence of the presence of mRNA for IL-4, IL-5 and IL-10 in tissues from DCL patients.[72] Although IL-2 and IFN-γ production in supernatant of lymphocyte cultures stimulated with leishmania antigen is absent, high levels of mRNA for IL-10 are found in mononuclear cells from DCL. More recently we observed that mRNA for IFN-γ is absent in peripheral blood mononuclear cells from five DCL patients with active disease and without therapy. In such cases mRNA for IL-10 and

IL-4 although present was quite variable and generally when the signal for IL-4 was strong, it was weak for IL-10 and vice-versa. After therapy and clinical improvment there was a decrease in the mRNA for IL-4 and IL-10 and mRNA for IFN-g was documented. Additionally, it was observed that anti-human IL-10 neutralizing monoclonal antibody restored lymphocyte proliferation in two out five DCL patients evaluated. The documentation of high mRNA for IL-10 in tissue and peripheral blood and capability of anti-IL-10 to restore in vitro T-cell function supports that IL-10 is participating in the down regulation of the cellular immune response in diffuse cutaneous leishmaniasis.

IL-10 IN CUTANEOUS AND MUCOSAL LEISHMANIASIS

Classical cutaneous leishmaniasis is characterized by well delineated ulcers with raised borders, although nodules, vegetation and acneiform lesions may also occur. Several species of leishmania may cause cutaneous ulcers. In endemic areas of *L. braziliensis,* and less frequently of *L. amazonensis,* transmission of mucosal disease occurs in about 3% of patients with cutaneous lesions.[73] Mucosal leishmaniasis is a severe and disfigurating disease and therapeutic failure and relapses are usually observed. Both cutaneous and mucosal leishmaniasis are associated with a strong T-cell response determined by IL-2 and IFN-γ production, as well as by delayed type hypersensitivity.[74,75] Cytotoxic cells are also found in peripheral blood of cutaneous and mucosal leishmaniasis patients. Such cells when stimulated by leishmania antigen are able to lyse both NK susceptible K562 and NK resistant Daudi tumor cells as well as autologous antigen pulsed macrophages.[76] In fact, these immunological responses to parasite antigens are quantitativelly higher in mucosal leishmaniasis than in cutaneous leish-maniasis.[74] In addition to T-cell function, macrophages from both cutaneous and mucosal leishmaniasis have in vitro capability to destroy amastigote forms of leishmania in the presence of superna-

tant of mucosal or cutaneous lymphocytes cultures stimulated with leishmania antigen.[74] Since the inflammatory reaction characterized by lymphocyte and plasmocyte infiltration is prominent in mucosal tissue and absence or few parasites are found in the lesion,[77] the possibility that mucosal damage is secondary to a hypersensitivity or autoimmune reaction has been proposed. The documentation that mucosal leishmaniasis patients produce great amounts of IFN-γ but they are not able to control the disease raises the possibility that in vivo macrophages from such disease have somehow controlled or suppressed their capability to destroy leishmania amastigotes.

Cytokine profiles (mRNA) from tissues of cutaneous and mucosal leishmaniasis patients have been determined using polymerase chain reaction. Comparative analysis has shown that while in cutaneous ulcers the cytokine pattern is of the Th1 type with the presence of IL-2, IFN-γ and lymphotoxin, in patients with mucosal leishmaniasis the pattern is of the Th0 type, with high amounts of IL-2, IFN-γ, but also of IL-4 and IL-10.[72] The cytokine pattern observed in tissue of mucosal leishmaniasis patients is also found in resting, or in leishmania antigen driven, peripheral blood mononuclear cells.[78] IL-10 has also been documented in supernatants of lymphocyte cultures stimulated with leishmania antigens. The reasons for the increase in mRNA and secretion of IL-10, and the cells involved in these phenomena are not completely understood. Macrophages, mast cells, B cells and Th2 cells are all capable of producing IL-10. The evidence of a Th0 pattern in mucosal leishmaniasis with documentation of IL-4, IL-5, IL-10 and IL-2, IFN-γ and lymphotoxin suggests that T cells are involved in IL-10 production in such cases. The possibility that the immunological response to specific leishmania antigens may induce predominantly the production of one type of cytokine rather than another is likely. In fact, peculiarities in the cytokine profile secreted by T cells from mucosal leishmaniasis patients are found in lymphocyte cultures stimulated by two different portions of the 83kd heat shock protein cloned from *L. braziliensis*.[78] Both Hsp83a and Hsp83b antigens induced IFN-γ and TNF-α production. However the Th2 cytokine (IL-10 and IL-4) profiles differed for both Hsp83 antigens. Lbhsp83a stimulated the production of IL-10 but not IL-4 mRNA while Lbhsp83b stimulated the production of IL-4 but not IL-10 mRNA.[78] This data indicates that an immunogenic T-cell antigen can be tailered to define epitope domains that may induce different cytokine profile.

In the in vitro response of mucosal leishmaniasis patients to crude leishmania antigen, high amounts of IL-2, IFN-γ, TNF-α, IL-4 and IL-10 are observed. These data are in agreement with the evidence of both strong cellular immune response detected by delayed type hypersensitivity IFN-γ production and cytotoxic activity against Daudi and K562 cells, as well as B-cell activation, documented by high levels of antibody to leishmania antigen. In vitro, both T-cell responses of mucosal patients associated with leishmania killing, such as macrophage activation and cytotoxicity, are dramatically suppressed by IL-10. Therefore, it is possible that the mucosal leishmaniasis is an example of human disease associated with activation of Th1 and Th2 cells but the effectiveness of Th1 function is suppressed by IL-10.

REFERENCES

1. Desjeux P. Human leishmaniasis: epidemiology and public health aspects. Wld Hlth Statist Quant 1992; 45:267-75.
2. Chang KP, Kong D, Bray RS. Biology of leishmania & leishmaniasis. In K.P. Chang & R.S. Bray, (Ed) Leishmaniasis. (1985) Elsevier, London, 1-30.
3. Sacks DL, Perkins PV. Identification of an infection stage of leishmania promastigotes. Science 1984; 223:1417-22.
4. Pearson RD, Steigbebel RT. Mechanisms of lethal effect of human serum upon leishmania donovani. J Immunol 1980; 125: 2195-2211.
5. Bryceson ADM, Preston PM, Bray RS et al. Experimental cutaneous leishmaniasis. II. Effects of immunosuppression and antigenic competition on the course of infection with Leishmania enrietti in the guinea-pig. Clin Exp Immunol 1972; 10:305-35.
6. Howard JG, Hale C, Liew FY. Immunological regulation of experimental cutaneous leishmaniasis. III. The nature and significance of specific suppression of cell-mediated immunity. J Exp Med 1980; 152:594-607.
7. Franke ED, McGreevy PB, Katz SP et al. Growth cycle-dependent generation of complement resistant leishmania promastigotes. J Immunol 1985; 134: 2713-19.
8. Puentes SM, Da Silva RP, Sacks DL et at. Serum resistance of metacydic stage leishmania major promastigote is due to release of C5b-9. J Immunol 1990; 145:4311-16.
9. Soares NM, Carvalho EM, Pinto RT et al. Induction of complement sensitivity in Leishmania amazonensis metacyclic promastigotes by protease treatment but not by specific antibodies. Parasitol Res 1993; 79:340-42.
10. Murray HW, Rabin BY, Rothermel CD. Killing of intracellular leishmania donovani by lymphokine stimulated human mononuclear phagocytes. Evidence that in IFN-g is the activating lymphokine. J Clin Invest 1983; 72:1506-10.
11. Nathan CF, Murray HW, Wiebe ME et al. Identification of IFN-γ as the lymphokine that activates human microbicidal oxidative metabolism and antimicrobicidal activity. J Exp Med 1983; 158:670-89.
12. Liew FY, Millett S, Parkinson C et al. Macrophage killing of leishmania parasite in vivo is mediated by nitric oxide from L. arginine. J Immunol 1990; 144:4794-97.
13. Chang KP, Chanduri G, Fong D. Molecular determinants of leishmania virulence. Annu Rev Microbiol 1990; 44:499-29.
14. Engelhorn S, Brucknery A, Remold HG. A soluble factor produced by inoculation of human monocytes with leishmania donovani suppresses γ Interferon dependent monocyte activation. J Immunol 1990; 145:2662-68.
15. Hall BF, Joiner KA. Strategies of obligate

intracellular parasites for evading host defenses. Immunol Today 1991; 12:A22-27.

16. Titus RG, Ribeiro JMC. Salivary gland lysates from the sandfly Lutzomya longipalpis enhance leishmania infectivity. Science 1988; 239:1306-08.

17. Theodos GM, Nong YH, Remold HG et al. Salivary gland material from the sandfly Lutzomya longipalpis has an inhibitory effect on macrophage function in vitro. Parasite Immunol, in press.

18. Carvalho EM, Barral A, Pedral-Sampaio D et al. Immunological markers of clinical evolution in children recently infected with Leishmania donovani chagasi. J Infect Dis 1992; 165:535-40.

19. Milon G, Titus RG, Cerottini JC et al. High sequency of leishmania major specific L3T4+ T cells in susceptible BALB/c mice as compared to resistant CBA mice. J Immunol 1986; 136:1467-73.

20. Farrel JP, Muller I, Louis JA. A role of Lyt -2+ T cells in resistance to cutaneous leishmaniasis in immunized mice. J Immunol 1989; 142:2052-59.

21. Hill JO, Awwad M, North RJ. Elimination of CD4 suppressor T cell from susceptible BALB/c mice releases CD8+ lymphocytes to mediate protective immunity against leishmania. J Exp Med 1989; 169:181-27.

22. Conceição-Silva F, Bertho A, Nogueira R et al. Leishmania braziliensis reactive T cells populations derived from peripheral blood of human American cutaneous leishmaniasis. CD4/CD8 ratio determined by monoclonal antibodies using the flow cytometry analysis. Mem Inst Oswaldo Cruz 1990; 84:92-8.

23. Scott P, Pearce E, Cheever AW et al. The role of T cell subsets and cytokines in the regulation of injection. Immunology Today 1991; 346-48.

24. Heinzel FP, Sadick MD, Mutha SS et al. Production of IFN-g, IL-2, IL-4 and IL-10 by CD4+ lymphocytes in vivo during healing and progressive immune leishmaniasis. Proc Natl Acad Sci USA 1991; 80:7011-15.

25. Heinzel FP, Sadick MD, Holaday BJ et al. Reciprocal expression of IFN-g or IL-4 during resolution or progression of murine leishmaniasis: evidence for expansion of distinct helper T cell subsets. J Exp Med 1989;

169:59-72.

26. Holaday BJ, Sadick MD, Wang Z et al. Reconstitution of leishmania immunity in severe combined immunodeficient mice using Th1 and Th2 like lines. J Immunol 1991; 147:1653-58.

27. Liew FY, Millott S, Schimitt JA. A repetitive peptide of leishmania can activate T helper type 2 cells and enhance disease progression. J Exp Med 1989; 172:1359-65.

28. Barral A, Barral-Netto M, Yong EL et al. Transforming growth factor b as a virulence mechanism for *Leishmania braziliensis*. Proc Natl Acad Science 1993; 90:3442-44.

29. Carvalho EM, Bacellar O, Brownell C et al. Restoration of lymphocyte proliferation and g-IFN production in visceral leishmaniasis. J Immunol 1994; 152:5949-56.

30. Barral-Netto M, Barral A, Brownell CE et al. Transforming growth factor-b in leishmanial infection: a parasite scape mechanism. Science 257:545-48.

31. Sher A, Coffman RL. Regulation of immunity to parasites by T cells and T cells derived cytokines. Ann Rev Immunol 1992; 10:385-91.

32. Silva JS, Morrissey PD, Grabstein KH et al. Interleukin-10 and IFN-g regulation of experimental *Trypanosoma cruzi* infection. J Exp Med 1992; 175:169-76.

33. Sieling PA, Abrams JS, Yamamura M et al. Immunosuppressive roles for IL-10 & IL-4 in human infection. In vitro modulation of T cell responses in leprosy. J Immunol 1993; 150:5501-10.

34. Clerici M, Wynn TA, Berzofsky JA, Blatt SP et al. Role of Interleukin-10 (IL-10) in T helper cell dysfunction in asymptomatic individuals infected with the human immunodeficiency virus (HIV-1). J Clin Inv 1994; 93:768-75.

35. Yssel H, de Waal Malefyt R, Roncarolo MG et al. IL-10 is produced by subsets of human CD4+ T cell clones and peripheral blood T cells1. J Immunol 1992; 149: 2378-84.

36. Moore KW, O'Garra A, de Waal Malefyt R et al. Interleukin-10. Ann Rev Immunol 1993; 11:165-90.

37. de Waal Malefyt R, Haanen J, Spits H et al. IL-10 and viral IL-10 strongly reduce

antigen specific human T cell proliferation by diminishing the antigen presenting capacity of monocytes via down-regulation of class II MHC expression. J Exp Med 1991; 174:915-21.

38. Macatonia SE, Doherty TM, Knight SC et al. Differential effect of IL-10 on dendritic cell-induced T cell proliferation and IFN-g production1. J Immunol 1993; 150: 3755-65.

39. Bogdan C, Vodovotz Y, Nathan C. Macrophage deactivation by Interleukin-10. J Exp Med 1991; 174:1549-60.

40. Gazzinelli RT, Oswald IP, James SL et al. IL-10 inhibits parasite killing and nitrogen oxide production by IFN-g activated macrophages. J Immunol 1992; 148:1792-96.

41. Taga K, Tosato G. IL-10 inhibit T cell proliferation and IL-2 production. J Immunol 1992; 148:1143-49.

42. Carvalho EM, Teixeira RS, Johnson Jr WD. Cell mediated immunity in American visceral leishmaniasis: reversible immunosuppression during acute infection. Infect Immun 1981; 33:498-02.

43. Carvalho EM, Bacellar O, Barral A et al. Antigen specific immunosuppression in visceral leishmaniasis is cell mediated. J Clin Invest 1989; 83:860-64.

44. Haldar JP, Ghose S, Saha KC et al. Cell mediated immune response in Indian kaleazar and post-kala-azar dermal leishmaniasis. Infect Immun 1983; 42:702-07.

45. Sacks DL, Lal SL, Shrivastava SN et al. An analysis of T cell responsiveness in Indian kala-azar. J Immunol 1987; 138:908-13.

46. Carvalho EM, Andrews BS, Martinelli R et al. Circulating immune complexes and rheumatoid factor in schistosomiasis and visceral leishmaniasis. Am J Trop Med Hyg 1983; 32:61-9.

47. Galvão-Castro B, Sá Ferreira GA, Marzochi KF et al. Polyclonal B cell activation circulating immune complexes and auto immunity in human American visceral leishmaniasis. Clin Exp Immunol 1984; 56:58-66.

48. Badaró R, Jones TC, Carvalho EM et al. New perspectives on a subclinical form of visceral leishmaniasis. J Infect Dis 1986; 154:1003-11.

49. Ho JL, Badaró R, Schwartz A et al. Diminished in vitro production of IL-1 and tumor necrosis factor-a during acute visceral leishmaniasis and recovery after therapy. J Infect Dis 1992; 165:1094-99.

50. Carvalho EM, Badaró R, Reed SG et al. Absence of g-IFN and Interleukin-2 production during active visceral leishmaniasis. J Clin Invest 1985; 76:2066-69.

51. Manson-Bahr PEC. Immunity in kala-azar. Trans Roy Soc Trop Med Hyg 1961; 55:550-55.

52. Andrade TM, Teixeira R, Andrade JAF et al. Estudo da hipersensibilidade do tipo retardado na leishmaniose visceral. Rev Inst Med Trop S Paulo 1982; 24:298-302.

53. Carvalho EM, Bacellar O, Reed SG et al. Visceral leishmaniasis: A disease associated with inability of lymphocytes to activate macrophages to kill leishmania. Braz J Med Res 1988; 21:85-92.

54. Zwingenberger K, Harms G, Pedrosa C et al. Determinants of the immune response in visceral leishmaniasis: Evidence for predominance of endogenous IL-4 over IFN-γ production. Clin Immunol Immunopathol 1990; 57:242-49.

55. Kemp M, Kurtzhals JAL, Bendtzen K et al. *Leishmania donovani* reactive Th1 and Th2 like T cells clones from individuals who have recovered from visceral leishmaniasis. Infect Immun 1993; 61:1069-75.

56. Ghalib HW, Piuvezam MR, Skeiky YAW et al. Interleukin 10 production correlates with pathology in human leishmania donovani infections. J Clin Invest 1993; 92:324-29.

57. Karp CL, El-Safi SH, Wynn TA et al. In vivo cytokine profiles in patients with kalaazar marked elevation of both IL-10 and g-IFN. J Clin Invest 1993; 91:1644-48.

58. Fiorentino D, Zlotnik F, Mosmann TR et al. IL-10 inhibits cytokine production by activated macrophages. J Immunol 1991; 147:3815-21.

59. de Waal Malefyt R, Abrams J, Bennett B et al. Interleukin-10 (IL-10) inhibit cytokine synthesis by human monocytes: An autoregulatory role of IL-10 produced by monocytes. J Exp Med 1991; 174:1209-20.

60. Reiner NE, Ng W, Ma T, McMaster WR.

Kinetics of g-interferon binding and induction of major histocompability complex class II mRNA levels in Leishmania-infected macrophages. Proc Natl Acad Sci USA 1988; 85:4330-34.

61. de Waal Malefyt R, Yssel H, de Vries JE. Direct effects of IL-10 on subsets of human CD4+ T cell clones and resting T cells. Specific inhibition of IL-2 production and proliferation1. J Immunol 1993; 150: 4754-65.

62. Fiorentino DF, Bond MW, Mosmann TR. Two types of mouse helper T cells. IV. TH2 clones secrete a factor that inhibits cytokine production by TH1 clones. J Exp Med 1989; 170:2081-98.

63. Powrie F, Menon S, Coffman RL. Interleukin-4 and Interleukin-10 synergize to inhibit cell-mediated immunity in vivo. Eur J Immunol 1993; 23:2223-29.

64. Afonso LCC, Scharton TM, Vieira LQ et al. The adjuvant effect of interleukin-12 in a vaccine against *Leishmania major*. Science 1994; 263:235-37.

65. Gazinelli RT, Hieny S, Wynn TA et al. Interleukin-12 is required for the T lymphocyte-independent induction of interferon-γ by an intracellular parasite and induces resistance in T cell-deficient hosts. Proc Natl Acad Sci USA 1993; 90:6115-19.

66. Locksley, RM. Interleukin-12 in host defense against microbial pathogens. Proc Natl Acad Sci USA 1993; 90:5879-80.

67. Bittencourt AL, Guimarães NA. Imunopatologia da leishmaniose tegumentar difusa. Med Cut ILA 1968; 2:395-402.

68. Bryceson ADM. Diffuse cutaneous leishmaniasis in Ethiopia I. The clinical and histopathological features of the disease. Trans Roy Soc Trop Med Hyg 1970; 63:708-37.

69. Convit J, Pinardi ME, Randon AJ. Diffuse

cutaneous leishmaniasis: a disease due to an immunological defect of the host. Trans Roy Soc Med Hyg 1972; 66:603-10.

70. Petersen EA, Neva FA, Oster CN et al. Specific inhibition of lymphocyte-proliferation response by adherent suppression cells in diffuse cutaneous leishmaniasis. New Eng J Med 1982; 306:387-91.

71. Petersen EA, Neva FA, Barral A et al. Monocyte suppression of antigen-specific lymphocyte responses in diffuse cutaneous leishmaniasis patients from the Dominican Republic. J Immunol 1984; 132:2603-06.

72. Dittmar G, Tapia FG, Sanchez MA et al. Determination of cytokine profile in American cutaneous leishmaniasis using Prolimerase Chain Reaction. Clin Exp Immunol 1993; 91:500-05.

73. Jones TC, Johnson WD, Barreto A et al. Epidemiology and clinical manifestation of American leishmaniasis due to L.*braziliensis*. J Infect Dis 1986; 156:73-83.

74. Carvalho EM, Johnson Jr WD, Barreto A et al. Cell mediated immunity in American cutaneous and mucosal leishmaniasis. J. Immunol. 135:4144-48.

75. Castes M, Agnelli A, Verde O et al. Characterization of the cellular immune response in American cutaneous leishmaniasis. Clin Exp Immunol 1983; 79:221-26.

76. Barral-Netto M, Barral A, Brodskyn C et al. Cytotoxicity in human mucosal and cutaneous leishmaniasis. Parasite Immunology 1994; in press.

77. Bittencourt AL, Andrade ZA. Aspectos imunológicos da leishamiose cutâneo-mucosa. Hospital (Rio) 1967; 71:975-84.

78. Skeiky YAW, Benson DR, Guderian JA et al. Recombinant leishmania braziliensis Hsp83: A pentavalent stimulators of leishmaniasis patient PBMC responses. Maunscript in preparation.

CHAPTER 11

IL-10 AND HIV-1 REPLICATION

Hanneke Schuitemaker, Neeltje A. Kootstra, and Frank Miedema

INTRODUCTION

The human immunodeficiency virus type 1 (HIV-1) causes the acquired immunodeficiency syndrome which is characterized by a gradual depletion of CD4+ T cells. Understanding which factors determine the rate of progression of HIV-1 infection is one of the major goals of current AIDS pathogenesis research. Biological variability of HIV-1 may be one of the factors that contribute to the differential course of infection. Biological properties with respect to which HIV-1 isolates can differ are for instance syncytium inducing (SI) capacity, replication rate and cytotropism.[1-5] HIV-1 isolates recovered from peripheral blood mononuclear cells of asymptomatic subjects are able to grow in phytohaemagglutinin (PHA) stimulated primary blood lymphocytes (PBL) but cannot be transmitted to cell lines, replicate relatively slowly and are non-syncytium inducing.[5] Primary virus isolation from approximately 50% of patients with AIDS related complex (ARC) or AIDS on PHA stimulated PBL results in the isolation of T-cell-line-tropic SI isolates. SI variants generally emerge during the course of infection and eventually are detected in about half of infected individuals with advanced disease.[6-8]

Next to CD4+ T cells, CD4 expressing monocytes/macrophages are targets for HIV-1.[9,10] NSI variants, in general, are much more capable or establishing a productive infection in primary (in vitro) monocyte derived macrophages (MDM) as compared to SI HIV-1 variants.[11]

The asymptomatic stage of HIV-1 infection is characterized by the presence of only low frequencies of infected peripheral blood T cells, in peripheral blood the only targets for HIV-1,[12] harboring predominantly NSI macrophage-tropic HIV-1 variants. It has been suggested that these variants, by virtue of their macrophage-tropism, may persist despite host immune surveillance. In this phase of infection, when high titers of HIV-1

Interleukin-10, edited by Jan E. deVries and René de Waal Malefyt.
© 1995 R.G. Landes Company.

neutralizing antibody are present, virus transmission may be successful only during direct cell-to-cell contact, which may frequently occur between macrophages and T cells in the process of antigen presentation. The observation that peripheral blood CD4⁺ T cells, in the asymptomatic phase of infection, carry only macrophage-tropic HIV-1 variants is indeed indicative for recent infection of these T cells by progeny from HIV-1 infected macrophages. (Fig. 11.1) This also suggests that HIV-1 infected macrophages may form the viral reservoir. Indeed in tissues, macrophages are the predominant HIV-1 infected cells.[10]

Later on in infection when the host immune system is compromised, an increase in the frequency of infected cells due to a selective expansion of more T-cell tropic HIV-1 variants is observed. The frequency of T cells harboring macrophage-tropic HIV-1 variants remains constant during all stages of infection even after the emergence of SI variants[11,13,14] indicative for their importance in viral persistence.

HIV-1 replication in macrophages is dependent on the activation and differentiation state of the cell and may thus be regulated by several cytokines. Enhanced virus expression in macrophages cultured in the presence of IL-3,[15-17] low dose IL-4,[17] GM-CSF,[15,17] and TNF-α and-β[18] has been reported. Exposure of MDM to TGF-β,[19] IL-13,[20] IFN-γ[21] or high dose IL-4[17] results in inhibition of HIV-1 replication. The reported altered levels of some of these cytokines in HIV-1 infected individuals[22] may thus have important consequences for virus replication in vivo. The mechanism of action by which these cytokines regulate HIV-1 expression has not been elucidated yet. However, NF-kB mediated enhancement of virus replication due to IL-1, TNF-α or IL-3 treatment seems plausible.[17,23]

High dose IL-4 (50 ng/ml), probably by inducing terminal differentiation of monocytes,[24] inhibits proliferative capacity of MDM and concomitantly prevents productive HIV-1 infection.[17] For establishment of productive infection in MDM, next

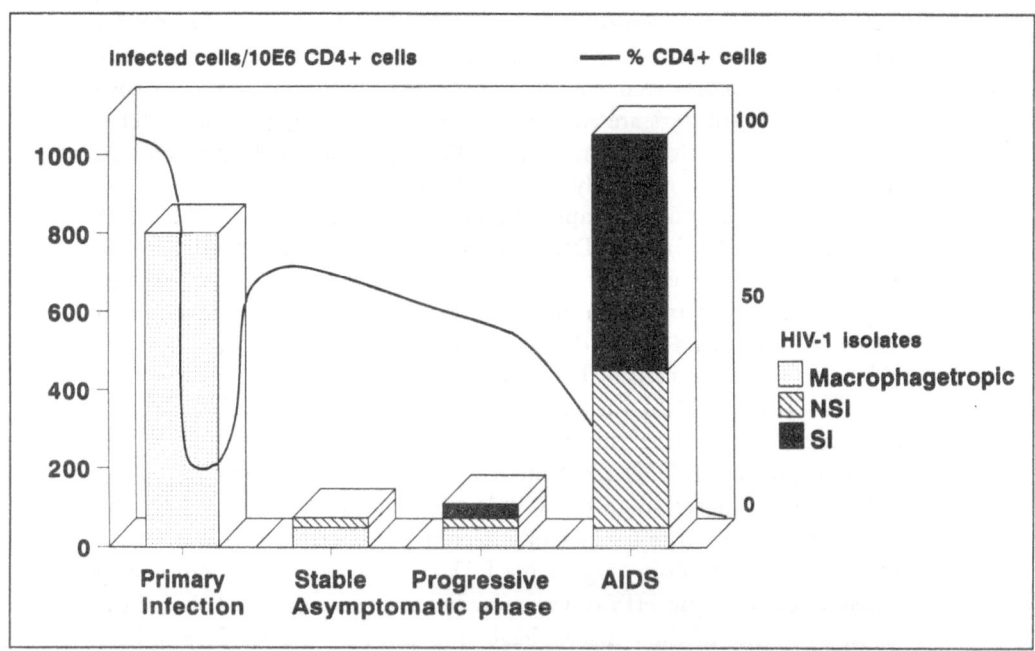

Fig. 11.1. Schematic representation of the prevalence of HIV-1 variants with different biological characteristics at distinct stages of HIV-1 infection. Left Y-axis: estimated proportion of HIV-1 infected cells per 10⁶ CD4⁺ cells; right Y-axis: percentage CD4⁺ cells. On X-axis, distinct stages in HIV-1 infection are depicted: Primary infection, stable to progressive asymptomatic infection, AIDS. Reprinted with permission from J Leukocyte Biol, 1994; 56 in press.

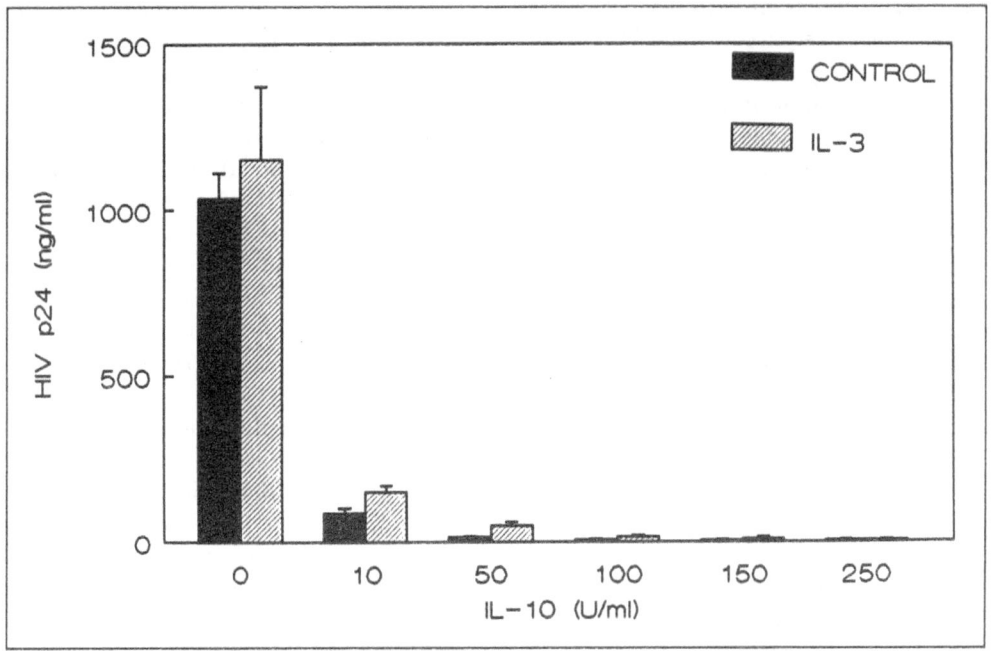

A

Fig. 11.2. Dose dependent inhibition of HIV-1 replication by IL-10.

(A.) Dose dependent complete inhibition of HIV-1 replication in MDM by IL-10. Cells were cultured for 5 days before inoculation in the presence of IL-3 (10 ng/ml) or medium alone in combination with increasing concentrations of IL-10. P24 production at day 14 is given. Results are the mean of four replicates and representative for three independent experiments.

(B.) Partial inhibition of HIV-1 replication in MDM by IL-10 treatment after inoculation. Cells were cultured for 5 days prior to HIV-1 inoculation in the presence of IL-3 (10 ng/ml) or medium alone. After inoculation at day 5 increasing concentrations of IL-10 were added to the cultures. P24 production at day 14 is given. Results are the mean of four replicates and representative for three independent experiments.

Reprinted with permission from J Virol et al. Interference of IL-10 with HIV-1 replication in primary monocyte derived macrophages. 1994; in press.

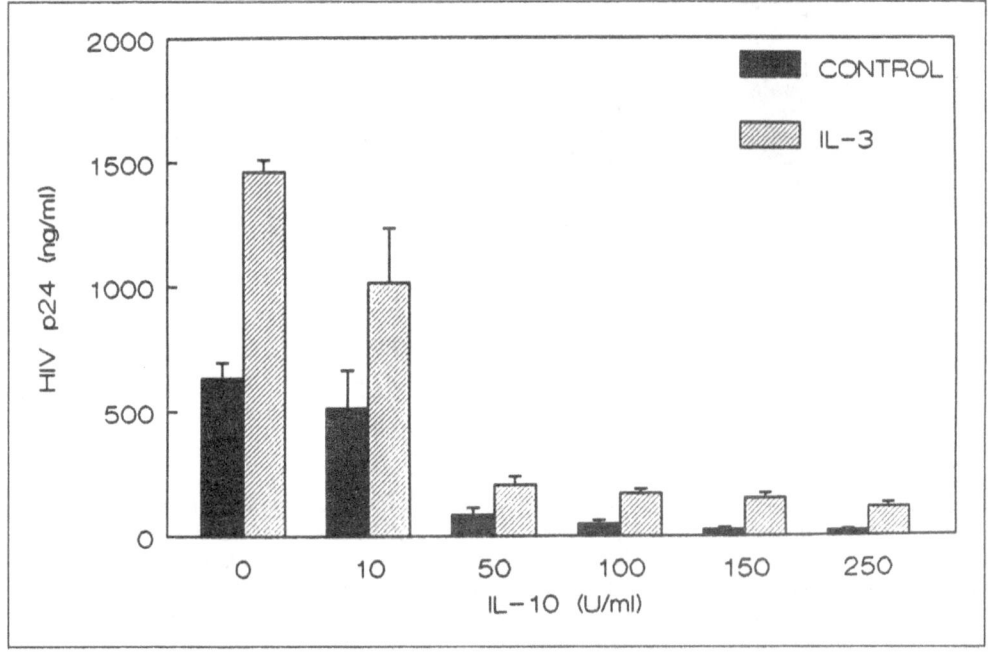

B

to a certain extent of differentiation, cell proliferation is required. Since HIV-1 is a retrovirus, it requires reverse transcription of its RNA genome into a proviral DNA species which can integrate in the host cell DNA. In non-proliferating MDM, the process of reverse transcription is disturbed resulting in only incomplete proviral DNA species.[17,25] Low dose IL-4 treatment does not affect proliferative capacity but gives rise to enhanced HIV-1 replication, most likely due to activation of NF-kB.[17]

IL-4 and IL-10 in many respects have common effects on macrophages. They both inhibit the production of IL-1α, IL-1β, IL-6, IL-8, TNFα, GM-CSF and G-CSF by monocytes at a transcriptional level.[26-28] These observations prompted us to study whether IL-10, in analogy to IL-4, had an effect on HIV-1 replication in MDM.[29] For this purpose monocytes were isolated from peripheral blood of HIV-1 seronegative plasmapheresis donors by Ficoll density gradient centrifugation followed by cen-

trifugal elutriation, as described previously.[30] Monocytes were subsequently exposed to increasing concentrations of IL-10 (range 0-250 U/ml) in medium alone or in combination with IL-3 (10 ng/ml) for 5 days prior to inoculation with the macrophage-tropic HIV-1 strain Ba-L.[10] Treatment with IL-10 before inoculation with HIV-1 resulted in a dose dependent inhibition of virus replication, which was complete at an IL-10 concentration of 100 U/ml (Fig. 11.2A). Addition of IL-10 after inoculation also resulted in a dose dependent inhibition of HIV-1 replication (Fig. 11.2B), but even in the presence of the highest dose of IL-10, complete inhibition of virus production could not be achieved.

MDM that had been precultured in the presence of IL-3 showed enhanced HIV-1 replication as compared to untreated controls (Fig. 11.2A), as has been previously reported.[15-17] However, MDM treated with both IL-3 and IL-10 prior to inoculation or MDM pretreated with IL-3 and cultured

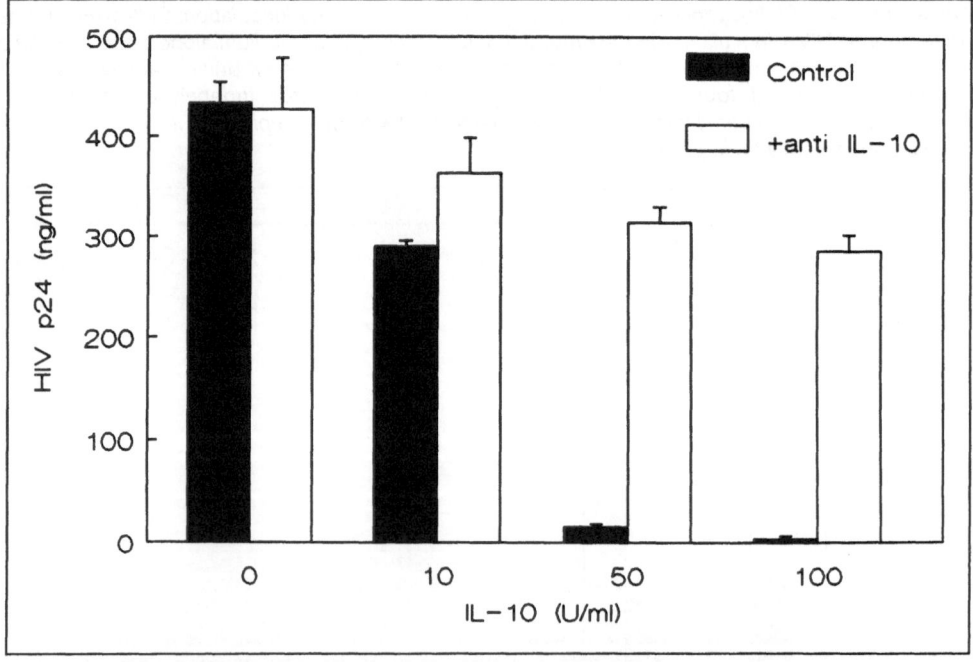

Fig. 11.3. Pre-incubation with an anti-IL-10 serum reverses IL-10 induced inhibition of HIV-1 replication in MDM. P24 production at day 14 is given. Results are the mean of 4 replicates and representative for three independent experiments. Closed bars represent p24 production of IL-10 treated cultures. Open bars represent p24 production of cultures in which IL-10 was pre-incubated with an anti IL-10 serum (5 µg/ml). Reprinted with permission from J Virol et al. Interference of IL-10 with HIV-1 replication in primary monocyte derived macrophages. 1994; in press.

in the presence of IL-10 after inoculation, had decreased virus production (Fig. 11.2 A and B). Complete abrogation of IL-10 mediated inhibition of virus replication by a neutralizing anti-IL-10 serum (5 μg/ml), confirmed that inhibition was specifically caused by IL-10 (Fig. 11.3).

Subsequently, the exact mechanism of action by which IL-10 interferes with HIV-1 replication was studied by investigating the influence of IL-10 on defined steps of the virus replication cycle (Fig. 11.4), such as entry and reverse transcription, RNA expression, and protein synthesis and processing. First, efficiency of reverse transcription as reflected by the presence of newly synthesized proviral DNA in HIV-1 inoculated IL-10 pretreated MDM was analyzed. PCR analysis was performed with a primer pair which specifically recognizes proviral DNA species that are synthesized relatively late in the process of reverse transcription. Presence of pro-viral DNA could be demonstrated (Fig. 11.5) when primer pairs were used that amplify the V3 domain of HIV-1 gp120 demonstrating not only normal viral entry but also efficient initiation of reverse transcription, irrespective IL-10 treatment. Inhibitory effects of IL-10 on the production of several cytokines by monocytes/macrophages have been reported.[26]

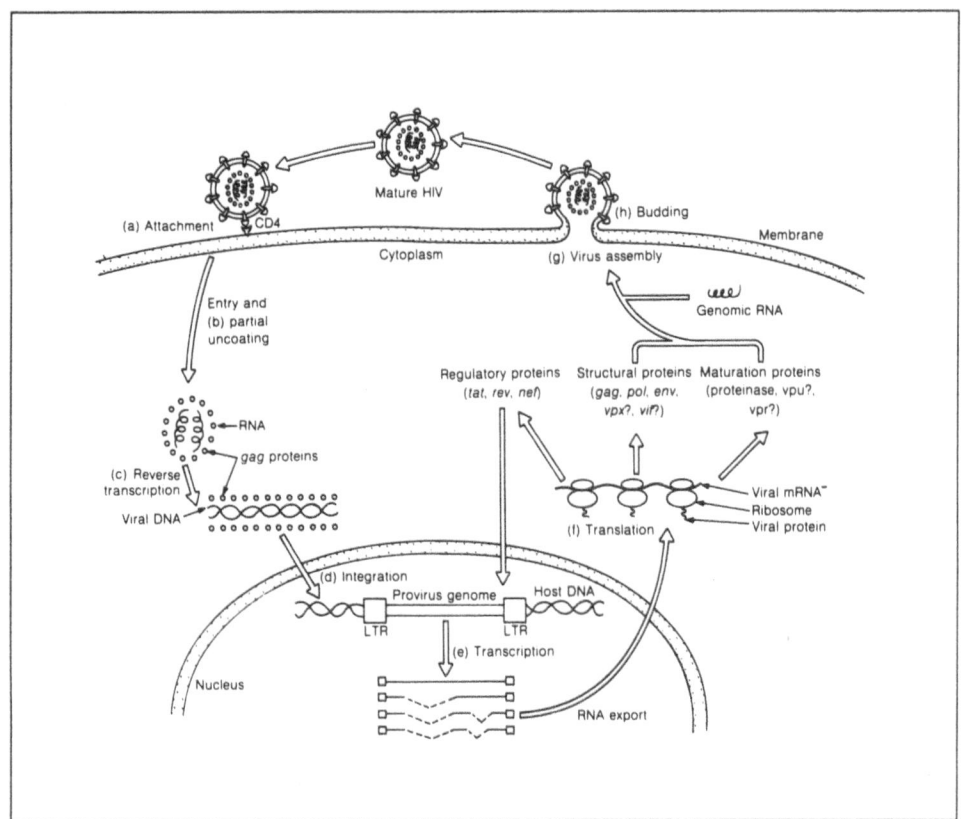

Fig. 11.4. Schematic representation of the HIV-1 replication cycle. (a) After attachment of virus particles to the CD4 receptor molecule, virus enters the cell. (b) The outer lipid envelope of the virus is removed when the particle undergoes fusion with the cytoplasmic vacuoles. (c) The core particle which remains is the site for reverse transcription of the virion RNA into DNA. (d) Following translocation into the nucleus, integration into the DNA of the cell occurs. (e) the integrated provirus is transcribed by cellular RNA polymerase II. (f) Translation of viral messenger (m) RNAs produces regulatory proteins which stimulate synthesis of structural proteins of the virion. (g) Accumulation of structural proteins in the cell membrane permits assembly of virus particles. (h) Maturation of virions and release from the cell by budding. Reprinted with permission from AIDS, 1989: suppl. 1, S19-S34.

Expression of these cytokines and HIV-1 are both predominantly regulated by NF-kB.[31-33] The IL-10 mediated inhibition of HIV-1 replication could thus be at a transcriptional level, associated with interference with NF-kB activation. The effect of IL-10 on HIV-1 transcription was first studied in a transient expression system. Monocytes were cultured for 5 days in the absence or presence of IL-3 (10 ng/ml) either or not in combination with IL-10 (100 U/ml). Cells were then transfected with CAT-constructs, in which the transcriptional activity was under control of either NF-kB linked to a heterologous promotor or the complete HIV-1 LTR. A modest stimulatory effect on both HIV-1 LTR and NF-kB driven CAT activity was found when MDM were treated with IL-10 alone or in combination with IL-3 (Fig. 11.6A), thus excluding that IL-10 mediated inhibition of virus replication occurred at a transcriptional level.

Possible effects of IL-10 treatment of monocytes on LTR-driven gene expression in MDM was further substantiated in Northern blot analysis (Fig. 11.6B). MDM were cultured in the absence or presence of IL-10 either before or after exposure to HIV-1 at day 5. Four days after HIV-1 inoculation RNA isolation and Northern blot analysis was performed. Hybridization was performed with a 3' LTR probe which recognizes unspliced, single and multiple spliced RNA transcripts. Irrespective IL-10 treatment, equal amounts of all HIV-1 specific mRNA species could be demonstrated (Fig. 11.6B). These results not only confirmed normal LTR driven HIV gene expression but also excluded an inhibitory effect of IL-10 on elongation of RNA transcription and splicing.

The normal expression of viral RNA products but absence of virus production in IL-10 treated MDM pointed to post-transcriptional interference by IL-10 with the HIV-1 replication cycle. Inhibition may occur during the process of virus assembly or budding, or at an earlier step, at the

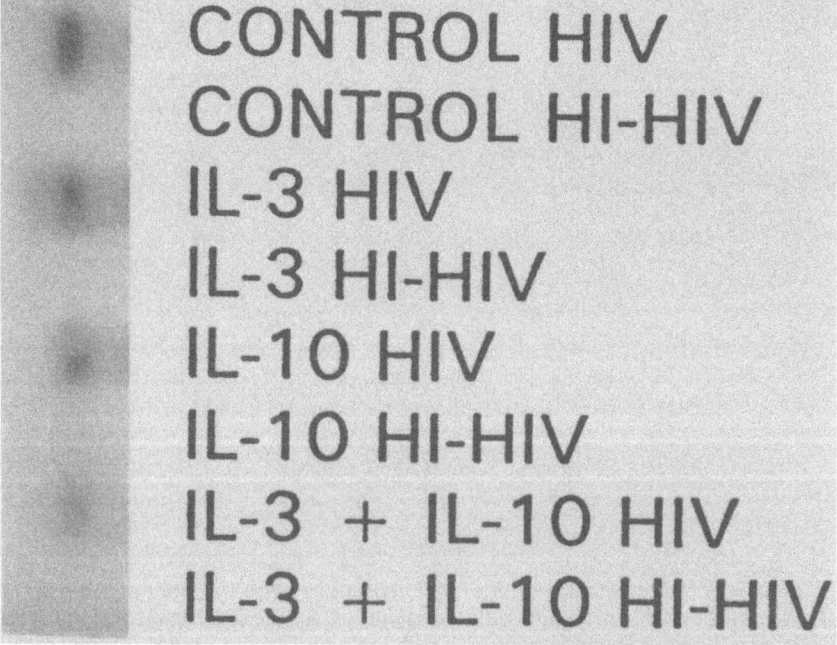

Fig. 11.5. Analysis on the presence of pro-viral DNA in IL-10 treated MDM. Amplification of the V3 region was successful irrespective IL-10 treatment prior to inoculation. MDM were cultured for 5 days prior to inoculation in medium, IL-3 (10 ng/ml), IL-10 (100 U/ml) or a combination of IL-3 and IL-10. The inoculum was DNase treated to distinguish for newly synthesized pro-viral DNA. As a negative control for PCR, a heat-inactivated inoculum (HI-HIV; 1 hour 56 °C) was used. DNA was extracted 48 hours after inoculation.

level of protein synthesis or processing. To analyze any inhibitory effect on the process of virus assembly or budding, MDM were cultured in the presence of IL-10 (100 U/ml) for 5 days prior to or directly after inoculation with HIV-1 Ba-L. Fourteen days after inoculation the presence of p24 was measured in culture supernatant and cell lysates. When MDM were treated with IL-10 before inoculation, p24 antigen could not be detected in culture supernatant nor in cell lysates. In contrast, although the amount of p24 antigen in the culture supernatant was very low, p24 antigen could be detected in cell lysates of MDM that had been treated with IL-10 directly after inoculation (Fig. 11.7). The intracellular accumulation of HIV-1 p24 antigen indeed pointed to an inhibitory effect at the level of virus assembly or budding in MDM that had been exposed to IL-10 treatment after establishment of infection.

In contrast, the absence of accumulated p24 but the presence of HIV-1 RNA species

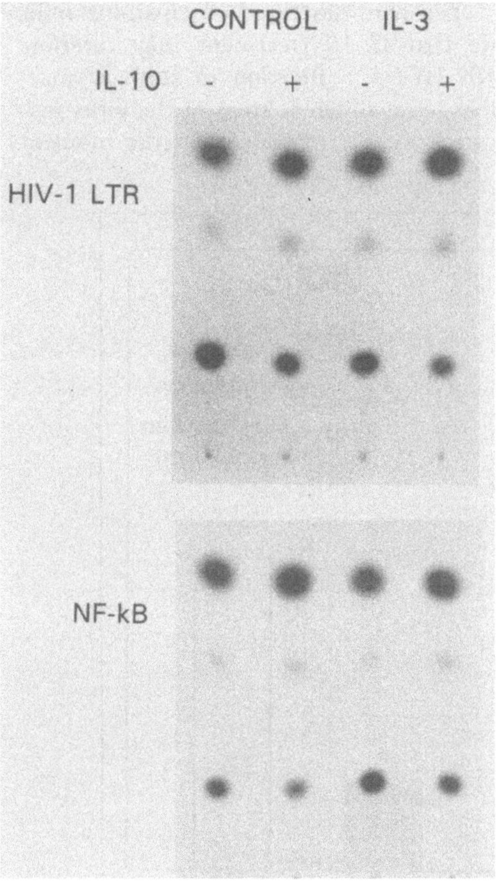

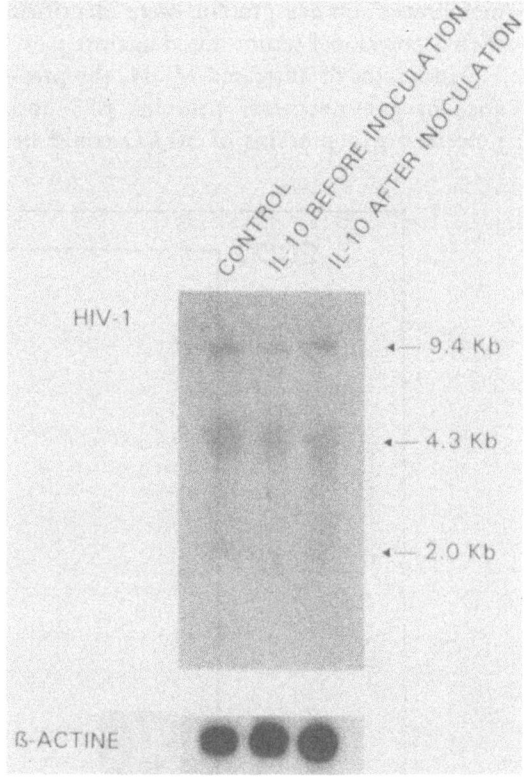

Fig. 11.6B. IL-10 does not interfere with HIV-1 replication at a transcriptional level. Absence of IL-10 mediated effect on HIV-1 specific mRNA expression as demonstrated by Northernblot analysis. MDM were cultured for 5 days prior to inoculation in medium with or without IL-3 (10 ng/ml) in the absence or presence of IL-10 (100 U/ml). After inoculation, cells were cultured in the absence or presence of IL-10 (100 U/ml) and total RNA was extracted after an additional 10 day culture. For the detection of HIV-1 specific mRNA transcripts, a 3' LTR probe was used. As a quantitative control for the total amount of RNA, a β-actine probe was used. Reprinted with permission from J Virol. Interference of IL-10 with HIV-1 replication in primary monocyte derived macrophages. 1994; in press.

Fig. 11.6A. IL-10 does not interfere with HIV-1 replication at a transcriptional level. Absence of an IL-10 mediated effect on HIV-1 LTR or NF-kB driven CAT-activity. MDM were cultured for 5 days before transfection in medium alone or in the presence of IL-3 (10 ng/ml). These pretreatments were performed in the absence or presence of IL-10 (100 U/ml). After 24 hours cell extracts were prepared for CAT assay. Results are representative for four independent experiments. Reprinted with permission from J Virol. Interference of IL-10 with HIV-1 replication in primary monocyte derived macrophages. 1994; in press.

in MDM that had been exposed to IL-10 prior to inoculation suggested an inhibitory effect on protein synthesis or processing in these cells. To analyze this possibility, the expression of viral proteins was studied in more detail on Immuno Blot. MDM were cultured in the presence or absence of IL-10 (100 U/ml) for 5 days prior to inoculation with HIV-1 Ba-L. At day 7 after HIV-1 inoculation cell lysates were prepared for analysis on a 3 mm 11% acrylamide gel in the presence of SDS. Proteins were transferred to nitrocellulose membranes and gag proteins were identified with a polyclonal serum raised against p24.

In lysates of untreated MDM, the presence of gag precursor proteins p53 and processed gag proteins of 30 kD could be

demonstrated. However, in lysates of MDM that had been cultured in the presence of IL-10 prior to inoculation, accumulation of gag precursor protein p53 could be demonstrated, but no evidence for the presence of processed gag proteins could be obtained (Fig. 11.8). This points to inhibition at the level of processing of viral proteins in MDM that were cultured in the presence of IL-10 prior to HIV-1 inoculation. IL-10 mediated inhibition of protease activity, which is required for gag processing, may be the underlying mechanism.

The here described observations indicate that IL-10 treatment may interfere with HIV-1 replication in primary macrophages at different steps of the virus replication cycle. Dependent on the moment

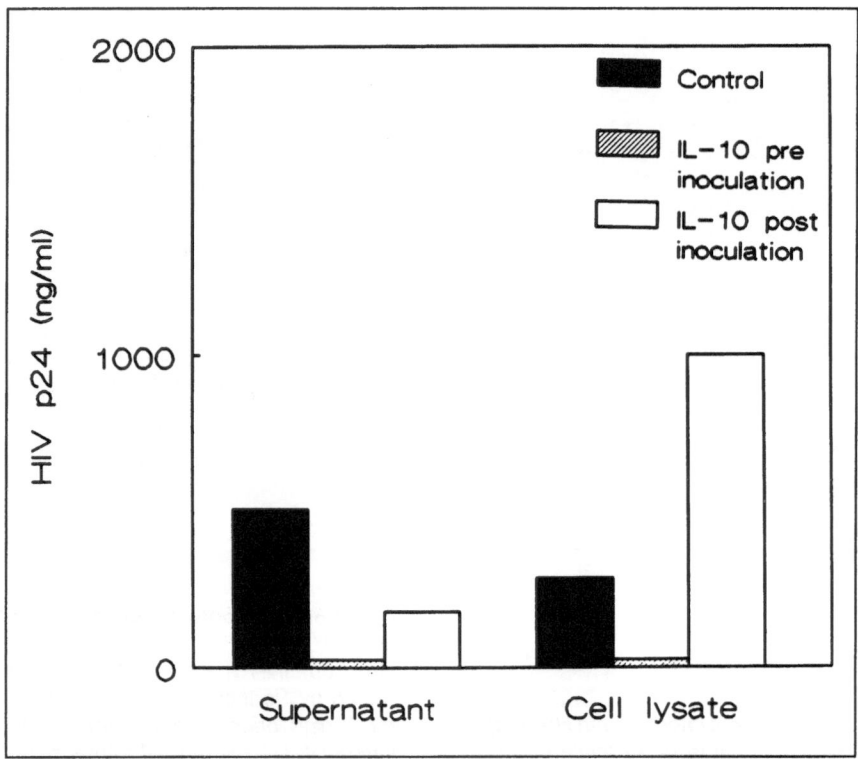

Fig. 11.7. Accumulation of p24 in MDM treated with IL-10 after inoculation. MDM were cultured for 5 days before inoculation in the presence or absence of IL-10 (100 U/ml), after inoculation unabsorbed virus was removed and the MDM were further cultured in the absence or presence of IL-10 (100 U/ml). 14 days after inoculation p24 production was measured in cell lysates and supernatant. Results are the mean of 3 replicates and representative for two independent experiments. Black bars: untreated control cultures; hatched bars: IL-10 treatment before inoculation; white bars: IL-10 treatment after inoculation.

of administration, thus prior to or after HIV-1 inoculation, IL-10 seems to interfere with protein processing or virus assembly, respectively. The monocyte differentiation state at the moment of IL-10 administration may be responsible for this difference. Inhibitory effects of IL-10 on cytokine expression in monocytes at a transcriptional level have been described.[26] In our current study, normal HIV-1 LTR driven expression as revealed by Northern blot analysis and in a transient expression system excluded an IL-10 mediated inhibitory effect on HIV-1 transcription.

Expression of TNF-α and IL-6, cytokines that may enhance HIV-1 replication in an autocrine fashion,[18,34] is down-regulated by IL-10.[26,35] Restricted TNF-α expression may thus be the cause of the observed IL-10 mediated inhibitory effects on HIV-1 replication in MDM. Exogenous TNF-α and IL-6 indeed restores HIV-1 replication in IL-10 treated MDM.[35] Interestingly, TNF-α and IL-6 reportedly enhance HIV-1 replication at a post-transcriptional level.[34] Thus our observation

that IL-10 interferes with HIV-1 replication at a post-transcriptional level is in agreement with the hypothesis that inhibition may be mediated by endogenous TNF-α.[35]

Decreased expression of cell surface molecules such as MHC class II and B7 due to IL-10 treatment was demonstrated but in these studies the mechanism of action was not revealed.[36,37] In our study, the presence of intracellular, but the absence of extracellular p24 antigen in MDM treated with IL-10 post HIV-1 inoculation, does not exclude a comparable inhibitory mechanism for the expression of the mentioned cell surface molecules and viral proteins. Further experiments are needed to establish the intracellular processes by which IL-10 inhibits HIV-1 production in MDM.

Several studies have pointed to altered T-cell cytokine production patterns, which may be relevant for AIDS pathogenesis. PBMC obtained from HIV-1 infected individuals early in infection are relatively normal with respect to cytokine produc-

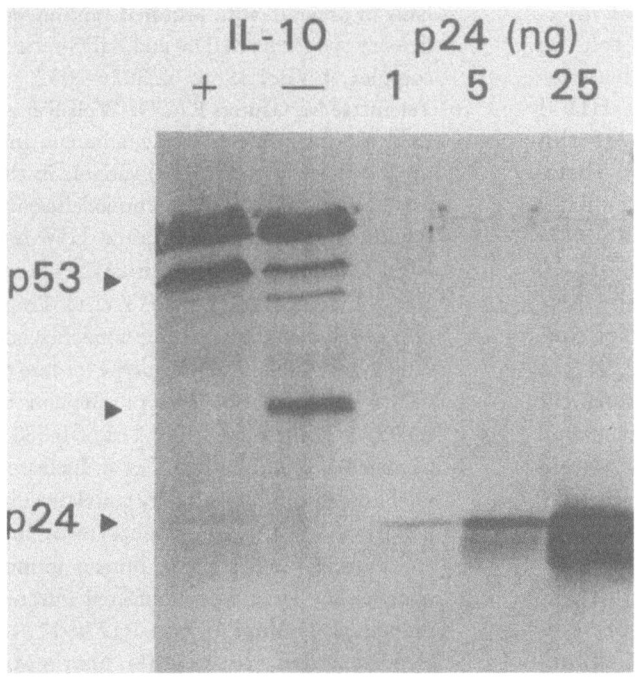

Fig. 11.8. Accumulation of p53 gag precursor protein, but absence of processed gag proteins in HIV-1 inoculated MDM cultured in the presence of IL-10 for five days prior to inoculation but normal expression of processed gag proteins in untreated MDM. MDM were cultured for 5 days prior to HIV-1 inoculation in the absence or presence of IL-10 (100 U/ml). At day 7 after inoculation, MDM were harvested and cell lysates were used for Immuno blot analysis. Reprinted with permission from J Virol. Interference of IL-10 with HIV-1 replication in primary monocyte derived macrophages. 1994; in press.

tion in response to recall antigens. In this asymptomatic stage, recall antigen stimulated PBMC predominantly produce IL-2. The production of IL-4 and IL-10 from these cells is low.[38] Later on in infection the cytokine production profile changes from predominantly IL-2 production in response to recall antigen towards an increased IL-4 and IL-10 production.[38-40]

In light of our current and previous[17] findings, elevated production of IL-4 and IL-10[38] may affect HIV-1 susceptibility of macrophages and HIV-1 replication in primary macrophages respectively. An increased susceptibility for HIV-1 of Th2 cells has been suggested (S. Romagnani, personal communication and A. Vyakarnam, personal communication). The shift from low frequencies of predominantly macrophage-tropic HIV-1 clones toward high frequencies of preferentially T-cell tropic clones, as we previously observed with progression of disease,[13] may be associated with this altered cytokine production profile and suggests an important contribution of cytokines to the regulation of HIV-1 tropism in vivo.

The presence of macrophage-tropic HIV-1 variants during all stages of infection has pointed to an important role for macrophages in viral persistence. Therefore, the inhibitory effect of IL-10 on HIV-1 replication in MDM would suggest this cytokine as a candidate for use in therapeutic intervention. Moreover, in contrast to the IL-4 mediated enhanced HIV-1 replication in peripheral blood T cells,[17] IL-10 had no effect on virus replication in these cells (data not shown). The important role, however, of both IL-4 and IL-10 in the cross regulation of Th phenotypes and the impaired immunity most likely associated with an expansion of Th2 cells[41] would argue against the use of these cytokines in HIV-1 infected individuals.

ACKNOWLEDGMENTS

The authors wish to thank Dr. Anthony Fauci for sharing unpublished results, Dr. Han Huisman for technical assistance, Angélique van 't Wout for helpful discussions and Linde Meyaard for critical reading of the manuscript. This work was supported by the Netherlands Ministry of Public Health (grants 88005 and 92024).

REFERENCES

1. Cheng-Mayer C, Seto D, Tateno M et al. Biologic features of HIV-1 that correlate with virulence in the host. Science 1988; 240:80-82.
2. Asjo B, Albert J, Karlsson A et al. Replicative properties of human immunodeficiency virus from patients with varying severity of HIV infection. Lancet 1986; ii:660-662.
3. Von Briesen H, Becker WB, Henco K et al. Isolation frequency and growth properties of HIV variants: multiple simultaneous variants in a patient demonstrated by molecular cloning. J Med Virol 1987; 23: 51-66.
4. Evans LA, McHugh TM, Stites DP et al. Differential ability of HIV isolates to productively infect human cells. J Immunol 1987; 138:3415-3418.
5. Tersmette M, De Goede REY, Al BJM et al. Differential syncytium-inducing capacity of human immunodeficiency virus isolates: frequent detection of syncytium-inducing isolates in patients with acquired immunodeficiency syndrome (AIDS) and AIDS-related complex. J Virol 1988; 62:2026-2032.
6. Tersmette M, Gruters RA, De Wolf F et al. Evidence for a role of virulent human immunodeficiency virus (HIV) variants in the pathogenesis of acquired immunodeficiency syndrome: studies on sequential HIV isolates. J Virol 1989; 63:2118-2125.
7. Koot M, Keet IPM, Vos AHV et al. Prognostic value of human immunodeficiency virus type 1 biological phenotype for rate of CD4+ cell depletion and progression to AIDS. Ann Int Med 1993; 118:681-688.
8. Connor RI, Mohri H, Cao Y et al. Increased viral burden and cytopathicity correlate temporally with CD4+ T-lymphocyte decline and clinical progression in human immunodeficiency virus type 1 infected infected individuals. J Virol 1993; 67:1772-1777.
9. Gendelman HE, Orenstein JM, Martin MA et al. Efficient isolation and propagation of human immunodeficiency virus on recom-

binant colony-stimulating factor 1-treated monocytes. J Exp Med 1988; 167: 1428-1441.

10. Gartner S, Markovits P, Markovits DM et al. The role of mononuclear phagocytes in HTLV-III/LAV infection. Science 1986; 233:215-219.

11. Schuitemaker H, Kootstra NA, De Goede REY et al. Monocytotropic Human Immunodeficiency Virus 1 (HIV-1) variants detectable in all stages of HIV infection lack T-cell line tropism and syncytium-inducing ability in primary T-cell culture. J Virol 1991; 65:356-363.

12. Schnittman SM, Psallidopoulos MC, Lane HC et al. The reservoir for HIV-1 in human peripheral blood is a T-cell that maintains expression of CD4. Science 1989; 245:305-308.

13. Schuitemaker H, Koot M, Kootstra NA et al. Biological phenotype of human immunodeficiency virus type 1 clones at different stages of infection: progression of disease is associated with a shift from monocytotropic to T-cell-tropic virus populations. J Virol 1992; 66:1354-1360.

14. Meltzer MS, Skillman DR, Hoover DL et al. Macrophages and the human immunodeficiency virus. Immunol. Today 1990; 11:217-223.

15. Koyanagi Y, O'Brien WA, Zhao JQ et al. Cytokines alter production of HIV-1 from primary mononuclear phagocytes. Science 1988; 241:1673-1675.

16. Schuitemaker H, Kootstra NA, Van Oers M et al. Induction of monocyte proliferation and HIV expression by IL-3 does not interfere with anti-viral activity of zidovudine. Blood 1990; 76:1490-1493.

17. Schuitemaker H, Kootstra NA, Koppelman MHGM et al. Proliferation dependent HIV-1 infection of monocytes occurs during differentiation into macrophages. J Clin Invest 1992; 89:1154-1160.

18. Poli G, Kinter A, Justement JS et al. Tumor necrosis factor alpha functions in an autocrine manner in the induction of human immunodeficiency virus expression. Proc Natl Acad Sci USA 1990; 87:782-785.

19. Poli G, Kinter AL, Justement JS et al. Transforming growth factor ß suppresses human immunodeficiency virus expression and replication in infected cells of the monocyte/macrophage lineage. J Exp Med 1991; 173:589-597.

20. Montaner LJ, Doyle AG, Collin M et al. Interleukin 13 inhibits Human Immunodeficiency Virus Type 1 production in primary blood-derived human macrophages in vitro. J Exp Med 1993; 178:743-747.

21. Kornbluth RS, Oh PS, Munis JR et al. Interferons and bacterial lipopolysaccharide protect macrophages from productive infection by human immunodeficiency virus in vitro. J Exp Med 1989; 169:1137-1151.

22. Vyakarnam A, Matear P, Meager A et al. Altered production of tumour necrosis factors alpha and beta and interferon gamma by HIV-infected individuals. Clin Exp Immunol 1991; 84: 109-115.

23. Osborn L, Kunkel S, Nabel GJ. Tumor necrosis factor a and interleukin 1 stimulate the human immunodeficiency virus enhancer by activation of the nuclear factor kappaB. Proc Natl Acad Sci. USA 1989; 86:2336-2340.

24. Te Velde AA, Klomp JPG, Yard BA et al. Modulation of phenotypic and functional properties of human peripheral blood monocytes by IL-4. J Immunol 1988; 140:1548-1554.

25. Schuitemaker H, Kootstra NA, Fouchier RAM et al. Productive HIV-1 infection of macrophages restricted to cell fraction with proliferative capacity. Submitted 1994.

26. de Waal Malefyt R, Abrams J, Bennett B et al. Interleukin-10(IL-10) inhibits cytokine synthesis by human monocytes: An autoregulatory role of IL-10 produced by monocytes. J Exp Med 1991; 174:1209-1220.

27. Te Velde AA, Huijbens RJF, Heije K et al. Interleukin-4 (IL-4) inhibits secretion of IL-1ß, tumor necrosis factor α, and IL-6 by human monocytes. Blood 1990; 76: 1392-1397.

28. Peleman R, Wu J, Fargeas C et al. Recombinant Interleukin-4 suppresses the production of Interferon-gamma by human mononuclear cells. J Exp Med 1989; 170:1751-1756.

29. Kootstra NA, Van 't Wout AB, Huisman JG et al. Interference of IL-10 with HIV-1

replication in primary monocyte derived macrophages. 1994; in press.

30. Figdor CG, Bont WS, Touw I et al. Isolation of functionally different human monocytes by counter-flow centrifugation elutriation. Blood 1982; 60:46-54.

31. Lenardo MJ, Baltimore D. NF-kB: A pleiotropic mediator of inducible and tissue-specific gene control. Cell 1989; 58:227-229.

32. Shakhov AN, Collart MA, Vassalli P et al. kB-type enhancers are involved in lipopolysaccharide-mediated transcriptional activation of the tumor necrosis factor a gene in primary macrophages. J Exp Med 1990; 171:35-47.

33. Nabel G, Baltimore D. An inducible transcription factor activates expression of human immunodeficiency virus in T cells. Nature 1987; 326:711-713.

34. Poli G, Bressler P, Kinter A et al. Interleukin 6 induces human immunodeficiency virus expression in infected monocytic cells alone and in synergy with tumor necrosis factor α by transcriptional and post-transcriptional mechanisms. J Exp Med 1990; 172:151-158.

35. Poli G, Weissman D, Kinter AL et al. Complex regulation of HIV replication by IL-4 and IL-10. J.Cell.Biochem. 1994; Suppl.18B:Abstr J261.

36. de Waal Malefyt R, Haanen J, Spits H et al. Interleukin-10 (IL-10) and viral IL-10 strongly reduce antigen-specific human T cell proliferation by diminishing the antigen-presenting capacity of monocytes via downregulation of class II Major Histocompatibility Complex expression. J Exp Med 1991; 174:915-924.

37. Ding L, Linsley PS, Huang L-Y et al. IL-10 inhibits macrophage costimulatory activity by selectively inhibiting the up-regulation of B7 expression. J Immunol 1993; 151:1224-1234.

38. Clerici M, Hakim FT, Venzon DJ et al. Changes in interleukin-2 and interleukin-4 production in asymptomatic human immunodeficiency virus-seropositive individuals. J Clin Invest 1993; 91:759-765.

39. Clerici M, Wynn TA, Berzofsky JA et al. Role of interleukin-10 (IL-10) in T helper cell dysfunction in asymptomatic individuals infected with the human immunodeficiency virus (HIV-1). J Clin Invest 1994; 93:768-75.

40. Meynard L, Otto SA, Keet IPM, Van Lier RAW, Miedema F. Changes in cytokine secretion patterns of CD4+ T cell clones in HIV-1 infection. Blood 1993; in press.

41. Meyaard L, Schuitemaker H, Miedema F. T-cell dysfunction in HIV infection: Anergy due to defective antigen presenting cell function? Immunol Today 1993; 14: 161-164.

CHAPTER 12

INTERLEUKIN-10 AND TRANSPLANTATION TOLERANCE

Maria-Grazia Roncarolo

INTRODUCTION

Despite significant progress in the past 20 years, induction of tolerance to host and donor HLA antigens after allogeneic transplantation, remains a rare event. Graft rejection and graft-versus-host disease (GVHD) are still the major obstacles to reach long-term engraftment and disease-free survival in patients transplanted with allogeneic hematopoietic stem cells, or solid organs.[1,2] Allograft rejection and GVHD are mainly due to activation of alloreactive T cells of host and donor origin, respectively. The recognition of alloantigens presented by either allogeneic, or autologous antigen-presenting cells (APC) results in activation of these allogeneic T cells.[3] This initial immune response leads to a cascade of events, including up-regulation of adhesion molecules, activation of macrophages, migration of T cells and recruitment of other effector cells, which cause tissue damage and the appearance of clinical manifestations of graft rejection and GVHD.[4,5] A large body of evidence supports the notion that cytokines play a critical role in these events. In particular, IL-2 and IFN-γ produced by the alloreactive T cells are involved in the initial amplification of the allogeneic responses, whereas IL-1 and TNF-α, produced by the APC, significantly contribute to the strong inflammatory responses which are characteristic for these diseases.[5-7] Therefore, modulation of cytokine production may represent a valuable target for immunosuppression.

Interleukin-10 (IL-10) is produced by a variety of cells including T lymphocytes and monocytes.[8] Interestingly, in contrast to murine IL-10, which is a Th2 product, human IL-10 is produced by Th0, Th1 and Th2 cells, but the levels of IL-10 production by Th2 cells are generally somewhat higher than those produced by the Th0 or Th1 subsets.[9] IL-10 has been described to reduce antigen-specific proliferative responses of human T cells. This inhibitory activity is due both to a direct effect of

Interleukin-10, edited by Jan E. deVries and René de Waal Malefyt.
© 1995 R.G. Landes Company.

IL-10 on IL-2 synthesis by T cells, and to indirect effects of IL-10 on the APC.[10,11] Although the exact mechanism of action of IL-10 to suppress the antigen presenting and accessory cell functions of monocytes/macrophages has not been completely elucidated, we know that IL-10 inhibits the production of various monokines and that it downregulates class II HLA expression on human monocytes.[11,12] Furthermore, recent data showed that IL-10 also downregulates the expression of the costimulatory molecule B7 on activated mouse macrophages.[13]

In addition to its suppressor activity on T-cell proliferation, IL-10 has potent anti-inflammatory effects. IL-10 inhibits the generation of reactive nitrogen intermediates by mouse macrophages, which are effector molecules involved in the elimination of intracellular and extracellular parasites.[8] In addition, IL-10 suppresses the production of pro-inflammatory cytokines (TNF-α, IL-1α, IL-1β, IL-6), chemokines (IL-8, MIP-1α) and hematopoietic growth factors (G-CSF, GM-CSF) by activated human monocytes.[8,12] In contrast, the production of the IL-1 receptor antagonist, which has been shown to have anti-inflammatory activity, is upregulated by IL-10.[8]

Based on these biological activities, we anticipated that IL-10 might play an important regulatory role in the process of allograft rejection and GVHD.

EFFECTS OF IL-10 ON ALLOGENEIC RESPONSES IN VITRO

Similar to its inhibitory effects on T-cell proliferation in response to soluble antigens, IL-10 was found to strongly reduce the proliferation of human alloreactive cells in classical one-way primary mixed lymphocyte reaction (MLR) in which peripheral blood mononuclear cells (PBMC) from two unrelated HLA mismatched individuals were used as stimulator and responder cells. Such inhibitory effects were observed also when purified CD3+ T cells were used as responder and purified allogeneic monocytes or B cells as stimulator cells. Interestingly, increased proliferation of the responder T cells was observed, when primary MLR were carried out in the presence of neutralizing anti-IL-10 mAbs, indicating that endogenous IL-10 secreted in primary MLR suppresses proliferative responses in these cultures.[14]

The levels of cytokines produced in MLR, in which unfractionated allogeneic PBMC were used as responder and stimulator cells, were significantly reduced in the presence of exogenous IL-10. The amounts of IL-2, IFN-γ, TNF-α and GM-CSF were reduced 2-3 fold. Furthermore, the proportions of T cells expressing the activation molecules CD25 and HLA-DR were lower in cultures containing IL-10, as compared to control cultures. Therefore, IL-10 suppresses proliferative T-cell responses toward alloantigens, not only by inhibiting cytokine production, but also by preventing T-cell activation.[14]

IL-10 also inhibited the generation of allospecific cytotoxic T cells in primary MLR. The levels of cytotoxic activity against the allogeneic targets were significantly reduced in the IL-10 containing cultures and enhanced in cultures carried out in the presence of anti-IL-10 mAb. Whether these results are due to direct inhibitory effects of IL-10 on the differentiation and activation of CD8+ cytotoxic T-cell precursors, or to indirect effects of IL-10 on activation of CD4+ T cells that provide help for the generation of mature cytotoxic T cells remains to be determined. However, the observation that IL-10 was able to reduce the proliferative responses of purified CD8+ T cells argues in favor of a direct effect.[14]

The proliferative responses of CD4+ allogeneic T-cell clones in vitro were also strongly suppressed by exogenous IL-10, when monocytes were used as stimulators. In parallel with the reduced proliferation, a reduction in the levels of IL-2, IL-5, GM-CSF and IFN-γ production by these T-cell clones was observed. Although significant amounts of IL-10, produced by monocytes were detected in these cultures, incubation with anti-IL-10 mAb did not significantly increase the proliferative responses and the cytokine synthesis by these T-cell clones (M. G. Roncarolo, unpublished data).[15] This may be due to the relative late production of IL-10 following activation of the T-cell clones and the requirement for IL-10 to act early in supporting T-cell proliferation.[9,14]

T-cell proliferation in primary MLR in which human dendritic cells were used as stimulators was also inhibited by IL-10.[16,17] The responses of purified CD4[+] and CD8[+] T cells to alloantigens presented by human skin derived Langerhans cells (LC) were strongly suppressed by IL-10.[16] In this study, IL-10 was found to act by reducing the antigen presenting capacity of freshly isolated, or cultured LC. The inhibitory effects of IL-10 could be partially overcome by exogenous IL-1, suggesting that suppression of IL-1 production by LC, at least in part, accounts for the inhibitory effects of IL-10.[16] Finally, IL-10 reduced the alloantigen presenting capacity of dendritic cells generated in cultures initiated with CD34[+] hematopoietic progenitor cells in the presence of GM-CSF and TNF-α.[17] Overall, these data indicate that IL-10 inhibits the responses of human T cells to alloantigens presented by professional APC.

IL-10 has similar effects on murine T cells. It inhibits the allogeneic responses of murine T cells to minor and major MHC antigens. Addition of IL-10 to MLR profoundly decreased the proliferation and IL-2 production by donor B10.BR cells stimulated with CBA cells which express different minor histocompatibility antigens (J. Ferrara, personal communication). IL-10 also strongly suppressed the proliferation of CD4[+] T cells, obtained from spleens of naive CBA/J mice, after stimulation with GM-CSF treated macrophages obtained from the bone marrow of BALB/c mice.[18] In this study, no inhibition of IL-2 production and T-cell proliferation was observed when purified dendritic cells were used as stimulators, which is in contrast with the data reported for human dendritic cells. However, IL-10 significantly inhibited dendritic cell-induced IFN-γ production by CD4[+] and CD8[+] T cells in a primary MLR.[18] The reason for these contrasting results obtained with human versus murine dendritic cells remains to be determined.

ROLE OF IL-10 IN TRANSPLANTED PATIENTS

SCID patients can be successfully transplanted with fetal hematopoietic stem cells derived from HLA mismatched donors. After immunological reconstitution, these children developed a chimerism, in which the T cells were of donor and the B cells and monocytes were of host origin.[19] Despite the HLA-incompatibility between the donor-derived T cells and the host cells, tolerance towards the donor and the host was established after transplantation, in the absence of any immuno-suppressive treatment. These patients had no signs of GVHD. In addition, the donor-derived T cells were specifically unresponsive to the host HLA antigens in primary MLR.[19] However, extensive studies performed with T-cell clones established from different transplanted patients, showed that the donor-derived T cells which recognize host HLA antigens were not deleted from the repertoire.[19,20] Both CD4[+] and CD8[+] T cells specific for the host class II and class I HLA antigens respectively, could be isolated with high frequencies from the peripheral blood of these patients several years after transplantation, indicating that these cells are not operational, and are suppressed in vivo. In general, the CD4[+] T cells displayed a low proliferative capacity and produced low levels of IL-2 after specific stimulation with host antigens. In contrast, high levels of IL-10 production were detected, particularly in one patient.[15] These high levels of IL-10 synthesis were only observed after stimulation via the TcR and were a specific property of the host-reactive clones. Alloreactive T-cell clones isolated from the same patient produced significant lower levels of IL-10, which were in the normal range, and comparable to those detected in the supernatants of alloreactive T-cell clones isolated from normal donors.

Kinetic studies indicated that T-cell clones isolated from normal donors produce IL-10 late after activation and no differences in these kinetics were observed between T-cell clones belonging to the Th0, Th1 or Th2 subsets.[9] In contrast, the host reactive T-cell clones produced IL-10 very early after activation. In addition, blocking experiments using neutralizing anti-IL-10 mAbs showed that this endogenously produced IL-10 was able to suppress the proliferation of these T-cell clones in an autocrine fashion.[15] In addition to the high IL-10 production by the host-reactive T-cell clones, high levels of IL-10 were also detected in the supernatants of PBMC of these patients cultured in vitro in the absence of any stimuli. These findings correlated with the high levels of IL-10 mRNA detected by semiquantitative PCR analysis of freshly isolated PBMC from SCID patients transplanted with fetal liver or with haploidentical bone marrow. Cell fractionation experiments indicated that, in addition to the donor-derived T cells, freshly isolated monocytes of host origin constitutively produced high levels of IL-10 in vivo.[15] This phenomenon was never observed with freshly isolated T cells or monocytes from normal donors. These data indicate that high levels of IL-10, produced by both host monocytes and donor-derived T cells, are present in patients in whom tolerance has been achieved after HLA-incompatible transplantation. The in vivo IL-10 production seems to be of biological relevance, since HLA-DR expression on monocytes of one of these patients was significantly lower as compared to monocytes isolated from normal donors. Furthermore, HLA-DR expression on the patient's monocytes could be partially upregulated by short term incubation of these cells with neutralizing anti-IL-10 mAbs. However, despite the high IL-10 levels in vivo, these patients are able to mediate normal primary and secondary immune responses to exogenous peptides.[15]

Collectively, these data indicate that endogenous IL-10 production is associated with tolerance in SCID patients transplanted with allogeneic stem cells and suggest that IL-10 may play an important role in establishing and maintaining transplantation tolerance, and also in preventing GVHD.

Although the mechanism of induction of endogenous IL-10 production remains to be determined, recent data indicated that high levels of spontaneous IL-10 production is a pre-existing condition in a subset of patients, who undergo stem cell transplantation. Preliminary analysis of spontaneous IL-10 production in vitro by PBMC of patients with hematological malignancies, prior to conditioning and bone marrow transplantation, showed that PBMC of patients with stable engraftment and high survival rates, produced higher levels of IL-10, as compared to the group of patients in whom severe GVHD or major complications developed after transplantation (E. Holler et al, manuscript in preparation). These data suggest that high levels of endogenous IL-10 prior to the transplant may set the stage for induction of tolerance and result in a favorable outcome of the process.

The role of IL-10 in preventing or suppressing rejection after solid organ transplantation is still under investigation. The presence of IL-10 has been analyzed in biopsies of patients who underwent heart or kidney transplantation[21,22] (P. Merville et al, manuscript submitted). The levels of IL-2, IL-4 and IL-10 mRNA in biopsies taken from hearts of patients with transplant rejection, were higher than those found in biopsies from hearts with no evidence of rejection. However, in contrast to IL-2, IL-10 was present only in heart biopsies of patients with mild rejections, and not in biopsies of patients with grade 3 rejection.[21] The conclusion of this study was that none of the cytokines evaluated, including IL-10, were predictive for rejection. Similarly, IL-10 secreting CD4[+] T cells have been detected at high frequencies among graft infiltrating lymphocytes obtained from biopsies of patients with acute rejection of kidney allografts. In contrast, low frequencies of CD4[+] IL-10 producing cells were observed in biopsies of patients with chronic rejection[22] (P. Merville et al, manuscript submitted). The major question which remains to be answered is whether endogenous IL-10 production contributed to the pathogenesis of the rejection, or whether it merely reflected an attempt of the immune system to counteract the high levels of IL-2, TNF-α, IL-1 produced by activated T cells and monocytes present at the site of rejection. Sequential analysis of in situ cytokine profiles during organ rejection in transplanted patients could help to address this question. Furthermore, it is of importance to elucidate the role of immunosuppressive therapy in influencing IL-10 production in these transplanted patients, particularly since it has been recently shown that cyclosporin A markedly enhances LPS-induced IL-10 release in vivo.[23] Therefore, it can not be ruled out that cyclosporin A treatment was responsible for the high levels of endogenous IL-10 production in some of these kidney- and heart-transplanted patients.

ROLE OF IL-10 IN ANIMAL MODELS OF TRANSPLANTATION

The role of IL-10 in murine models of organ rejection or graft-versus-host disease is still unclear and controversial. Endogenous IL-10 production has been detected in murine models of tolerance and GVHD. Differential activation of Th2-like effector cells has been reported by Takeuchi et al in a murine heart transplant model.[24] In this study, it was shown that IL-4 and IL-10 transcripts were generally enhanced in the heart grafts and spleens of mice rendered tolerant by donor-specific blood transfusions, anti-CD40 mAb pre-treatment and cyclosporin administration. In agreement with this study, Abramowicz et al reported that injection of semi-allogeneic spleen cells into neonatal Balb/c mice results in a state of tolerance, which is associated with in vivo expression of IL-4 and IL-10 mRNAs.[25] In addition, co-injection of Th2 cells into a B6C3F1 recipient prevents LPS-induced lethality during murine graft-versus-host reaction caused by injection of B6 spleen cells. These CD4+ Th2

cells were obtained from B6 mice treated in vivo with a combination of IL-2 and IL-4 and were shown to secrete enhanced levels of IL-4 and IL-10.[26] Taken together, these studies strongly suggest a protective role of IL-4 and IL-10 in organ rejection and GVHD. However, IL-10 does not always protect mice from organ graft rejection and GVHD. High levels of IL-10 mRNA have been detected in spleens of mice undergoing acute, or chronic GVHD.[27-29] Furthermore, no inhibition or delay in rejection was observed when IL-10 producing transgenic fetal pancreas, or adult islets were transplanted into MHC-incompatible recipients.[30] However, it remains to be determined whether the relatively high local levels of endogenous IL-10 plays a role in the pathogenesis of the GVHD and organ rejection.

Murine studies aimed at investigating the in vivo effects of IL-10 administration on GVHD have also generated controversial results. Preliminary experiments in GVHD models, in which the disease is induced by transplantation of CD4+ MHC Class II mismatched T cells into irradiated hosts, indicated that administration of IL-10 prior, and in the first days post-transplant increased the survival rate, but did not significantly reduce the GVHD symptoms, such as weight loss (E. Holler, R. Korngold, personal communication). IL-10 had no clear protective effects in models of minor MHC disparity mediated by CD4+ or CD8+ T cells. However, in some of these studies the in vitro IFN-γ production by T cells obtained from the treated animals was suppressed (J. Ferrara, personal communication). Finally, injection of IL-10 in a murine model of GVHD, due to both major and minor MHC disparities, failed to show any protection, but resulted in an exacerbation of the disease (B. Blazar et al, manuscript submitted).

Based on these data, it is difficult to draw any definitive conclusions on the role of IL-10 in murine organ rejection and GVHD. Further studies are warranted to understand the mechanisms and the conditions by which IL-10 induces protection from or exacerbation of GVHD and transplant rejection. In addition, the establishment of an in vivo human model of organ rejection in SCID-hu mice will help us to address these questions and to reconciliate the different results obtained in these murine models.

CONCLUDING REMARKS

Alloreactive T cells initiate the immune responses which lead to GVHD and transplant rejection. In addition, these processes have a strong inflammatory component, in which pro-inflammatory cytokines and chemokines produced by alloreactive T cells and monocytes/macrophages play an important role. IL-10 has strong anti-inflammatory activity. It inhibits the production of pro-inflammatory cytokines and chemokines, which are responsible for attracting inflammatory cells to the rejection site. On the other hand, IL-10 enhances the production of IL-1ra by monocytes, which by itself is an anti-inflammatory cytokine, because it blocks the activities of IL-1α and IL-β. In addition, IL-10 inhibits T-cell activation by soluble or alloantigens, either directly, by suppressing the IL-2 production, or indirectly, by reducing the antigen presenting and accessory function of APC. Based on these properties it has been proposed that IL-10 may function as a natural dampener of immune and inflammatory responses. In addition, as discussed here, IL-10 may have potential clinical utility as an immunosuppressive agent for preventing or reducing transplant rejection and establishing transplantation tolerance. Although it is still too early to draw firm conclusions, it is tempting to speculate that the high levels of endogenous IL-10 production observed in successfully transplanted SCID patients contribute to the tolerance achieved. On the other hand, it cannot presently be excluded that the high levels of endogenous IL-10 in the transplanted SCID patients were already present pre-transplantation. The presence of relatively high levels of endogenous IL-10 pre-transplant seems to be critical for successful allogenic bone marrow transplantation, since these patients have survival rates that are much higher than those of patients, who had low levels of endogenous IL-10 production.

Taken together, these data suggest that the levels of endogenous IL-10 may determine the outcome of GVHD and transplant rejection. Preliminary data indicating that GVHD in mice can be prevented by starting IL-10 administration prior to the transplant, also point in this direction. It has to be noted however, that IL-10 was not effective in other transplantation models. In some cases even exacerbation of GVHD was observed. In these situations IL-10 was given at the same time of the transplant, which may be too late to be effective. It is clear that more information is required, not only about the schedule of IL-10 administration, but also about optimal doses and duration of the IL-10 therapy. In addition, it is presently not clear whether IL-10 indeed can induce tolerance, and more importantly, whether IL-10 is required for the maintenance of tolerance. Studies addressing these questions will give us more information about the mode of action of IL-10 in transplantation tolerance.

ACKNOWLEDGMENTS

The author thanks J. E. de Vries, R. de Waal Malefyt and R. Bacchetta for helpful discussions and critical review of the manuscript, and Jo Ann Katheiser for secretarial assistance.

REFERENCES

1. Häyry P, Isoniemi H, Yilmaz S et al. Chronic allograft rejection. Immunological Reviews 1993; 134:33-81.
2. Parkman R. Human Graft-versus-host disease. Immunodeficiency Reviews 1991; 2:253-64.
3. Sayegh MH, Watschinger B, Carpenter CB. Mechanisms of T cell recognition of alloantigen. The role of peptides. Transplantation 1994; 57:1295-1302.
4. Wecker H, Auchincloss Jr H. Cellular mechanisms of rejection. Curr Opin Immunol 1992; 4:561-66.
5. Antin JH, Ferrara JLM. Cytokine dys-

regulation and acute graft-versus-host disease. Blood 1992; 12:2964-68.

6. Dallman MJ. Cytokines as mediators of organ graft rejection and tolerance. Curr Opin Immunol 1993; 5:788-93.

7. Sykes M. Novel approaches to the control of graft-versus-host disease. Curr Opin Immunol 1993; 5:774-81.

8. Moore KW, O'Garra A, de Waal Malefyt R et al. Interleukin-10. Ann Rev Immunol 1993; 11:165-90.

9. Yssel H, de Waal Malefyt R, Roncarolo MG et al. Interleukin-10 is produced by subsets of human CD4+ T cell clones and peripheral blood T cells. J Immunol 1992; 149:2378-84.

10. de Waal Malefyt R, Yssel H, de Vries JE. Direct effects of IL-10 on subsets of human CD4+ T cell clones and resting T cells. Specific inhibition of IL-2 production and proliferation. J Immunol 1993; 150: 4754-65.

11. de Waal Malefyt R, Haanen J, Spits H et al. IL-10 and viral IL-10 strongly reduce antigen specific human T cell proliferation by diminishing the antigen presenting capacity of monocytes via downregulation of class II MHC expression. J Exp Med 1991; 174:1209-20.

12. de Waal Malefyt R, Abrams J, Bennett B et al. IL-10 inhibits cytokine synthesis by human monocytes: An autoregulatory role of IL-10 produced by monocytes. J Exp Med 1991; 174:1209-20.

13. Ding L, Linsley PS, Huang L-Y, Germain RN et al. IL-10 inhibits macrophage costimulatory activity by selectively inhibiting the up-regulation of B7 expression. J Immunol 1993; 151:1224-34.

14. Bejarano MT, de Waal Malefyt R, Abrams JS et al. Interleukin-10 inhibits allogeneic proliferative and cytotoxic T cell responses generated in primary lymphocyte cultures. Int Immunol 1992; 4:1389-97.

15. Bacchetta R, Bigler M, Touraine J-L et al. High levels of IL-10 production in vivo are associated with tolerance in a SCID patient transplanted with HLA mismatched hematopoietic stem cells. J Exp Med 1994; 179:493-502.

16. Péguet-Navarro J, Moulon C, Caux C et al.

Interleukin-10 inhibits the primary allogenic T cell response to human epidermal Langerhans cells. Eur J Immunol 1994; 24:884-91.

17. Caux C, Massacrier C, Vanbervliet B et al. Interleukin-10 inhibits T cell alloreaction induced by human dendritic cells. Int Immunol 1994; in press.

18. Macatonia SE, Doherty TM, Knight SC et al. Differential effect of IL-10 on dendritic cell-induced T cell proliferation and IFN-g production. J Immunol 1993; 150:3755-65.

19. Bacchetta R, Vandekerckhove BAE, Touraine JT et al. Chimerism and tolerance to host and donor in severe combined immunodefieciences transplanted with fetal liver stem cells. J Clin Invest 1993; 91:1067-78.

20. Roncarolo MG, Yssel H, Touraine JL et al. Autoreactive T cell clones specific for class I and class II HLA antigens isolated from human chimera. J Exp Med 1988; 167: 1523-34.

21. Cunningham DA, Dunn MJ, Yacoub MH, Rose ML. Local production of cytokines in the human cardiac allograft. Transplantation 1994; 57:1333-37.

22. Merville P, Pouteil-Noble C, Wijdenes J et al. Detection of single cells secreting IFN-gamma, IL-6 and IL-10 in irreversibly rejected human kidney allografts and their modulation of IL-2 and IL-4. Transplantation 1993; 55:639-46.

23. Durez P, Abramowicz D, Gérard C et al. In vivo induction of Interleukin-10 by anti-CD3 monoclonal antibody or bacterial lipopolysaccharide: Differential modulation by cyclosporin A. J Exp Med 1993; 177: 551-55.

24. Takeuchi T, Lowry RP, Konieczny B. Heart allografts in murine systems. Transplantation 1992; 53:1281-94.

25. Abramowicz D, Durez P, Donckier GV et al. Neonatal induction of transplantation tolerance in mice is associated with in vivo expression of IL-4 and -10 mRNAs. Transplantation Proceedings 1993; 25:312-13.

26. Fowler DH, Kurasawa K, Husebekk A et al. Cells of Th2 cytokine phenotype prevent LPS-induced lethality during murine graft-versus-host reaction. J Immunol 1994; 152:1004-13.

27. Allen RD, Staley TA, Sidman CL. Differential cytokine expression in acute and chronic murine graft-versus-host diseases Eur J Immunol 1993; 23:333-37.

28. Garlisi CG, Pennline KJ, Smith SR et al. Cytokine gene expression in mice undergoing chronic graft-versus-host disease. Mol Immunol (England) 1993; 30:669-77.

29. De Wit D, Van Mechelen M, Zanin C et al. Preferential activation of Th2 cells in chronic graft-versus-host reaction. J Immunol; 1993 150:361-66.

30. Lee M-S, Wogensen L, Shizuru J et al. Pancreatic islet production of murine interleukin-10 does not inhibit immune-mediated tissue destruction. J Clin Invest 1994; 93:1332-38.

CHAPTER 13

EFFECTS OF IL-10 IN LIPOPOLYSACCHARIDE- AND SUPERANTIGEN-INDUCED LETHAL SHOCK IN VIVO

Albert Zlotnik and Rachel A. Freiberg

INTRODUCTION

Interleukin-10 (IL-10) is a cytokine that has generated significant excitement since its discovery.[1] This cytokine, as discussed elsewhere in this book, exhibits pleiotropic biological activities on various lineages of murine and human cells. These activities, summarized in Figure 13.1, include growth cofactor for some cells, (mast cells,[2] thymocytes,[3] etc.) as well as, differentiation effects (Ia induction in B cells).[4] However, the activity that has generated the most enthusiasm is its ability to downregulate macrophage function[5,6] in a variety of ways. The latter effects of IL-10 strongly suggest that it has anti-inflammatory and/or immunomodulatory effects.

IL-10 PREVENTS THE TOXIC EFFECTS OF LPS IN VIVO

In macrophages, IL-10 downregulates the expression of several cytokines[6] usually associated with inflammation (such as IL-1, IL-6 and TNF-α) (Fig. 13.2). This observation is especially important, since it suggested that IL-10 could have anti-inflammatory properties. The first formal demonstration of the anti-inflammatory effects of IL-10 in vivo came from its ability to prevent lipopolysaccharide (LPS)-induced toxic shock in mice.[7,8] LPS is generally responsible for septic shock in humans and represents a significant public health problem. Accordingly, much effort has been directed at controlling the inflammatory events that mediate this response. In this system, it had been shown that the main target for LPS is the macrophage, which is then activated to produce large amounts of some cytokines (IL-1, IL-6, TNF-α, as well as many chemokines). Experiments using neutralizing anti-cytokine antibodies have strongly suggested that TNF-α is the main mediator of toxic shock in this model. In fact, anti-TNF-α antibodies prevent, to a significant extent, the toxicity of LPS in vivo. Given the demonstrated in vitro ability of IL-10 to inhibit the activation of macrophages in response to LPS,[6] it was a good candidate to inhibit the toxicity of LPS in vivo as well.

The results of such an experiment indicated that IL-10 was indeed capable of preventing the toxic effects of LPS in vivo. IL-10 could be administered up to 0.5 hours after LPS and still exhibit a protective effect in mice.[7] Furthermore, measurements of TNF-α in the sera of these mice showed that the circulating levels of TNF-α in mice treated with IL-10 re-

mained well below those of untreated mice challenged with LPS.[7,8]

Two experimental protocols have been utilized to study LPS-induced toxicity. One relies on the administration of LPS alone, while the other includes pre-treatment of mice with D-galactosamine followed by LPS. The precise mechanism of action of D-galactosamine is not known, but it is believed to render the mice more susceptible to the toxic effects of LPS by neutralizing to a certain extent the ability of the liver to clear LPS from the circulation.[9] IL-10 has been shown to be effective in preventing the toxic effects of LPS in either model (whether the protocol includes D-galactosamine or not).

IL-10 INHIBITS T-CELL ACTIVATION IN VIVO

The results discussed above suggest a potential use for IL-10 in septic shock. However, there are other possibilities for therapeutic use of IL-10 derived from its known in vitro activities. The most important focuses, again, on its effects on macrophages. As shown in Figure 13.1, another

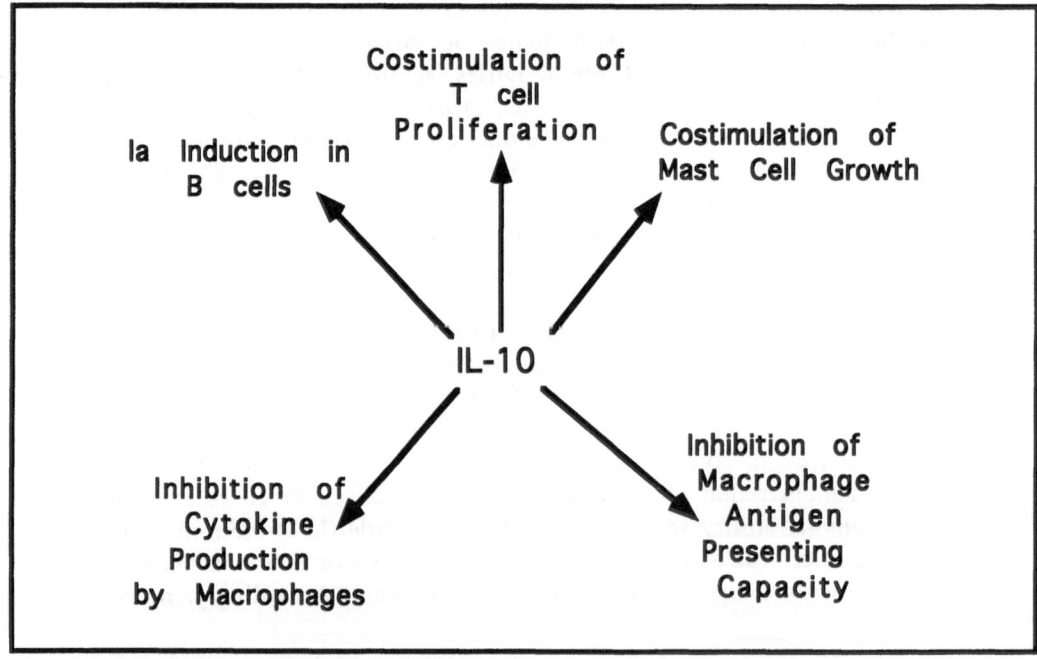

Fig. 13.1. Selected biological activities of IL-10.

macrophage function inhibited by IL-10 is the ability of these cells to present antigen and stimulate T-cell activation. As shown in Figure 13.3, in a simplified way, T-cell activation is a phenomenon that requires an antigen-presenting cell to process antigen and present it in an appropriate manner (usually as an immunogenic peptide) to T cells with the appropriate T-cell receptor specificity. As a secondary (but also pivotal) requirement, T cells require triggering through costimulatory molecules, of which many have been described (including CD28-B7/B70, LFA-1-ICAM-1, CD40-GP39, etc). The relative importance of these interactions probably depends on the nature of the antigen-presenting cell. Antigen-presenting cells must be class II MHC positive, and also express some of the appropriate secondary costimulatory molecules. The main antigen-presenting cells can therefore be dendritic cells, macrophages or B cells. As shown in Figure 13.1, IL-10 induces class II MHC expression in mouse B cells. This suggests that IL-10 could potentiate the ability of B cells to activate T cells. In contrast, IL-10 inhibits the ability of macrophages to activate T cells (as discussed

at length elsewhere in these reviews). It should be pointed out though, that the precise mechanism for this inhibition has not yet been elucidated. Finally, the effects of IL-10 on dendritic cells (an important, but less abundant type of antigen-presenting cell) are still being characterized. Given these in vitro observations, it was not easy to predict the potential of IL-10 to modulate T-cell activation in vivo, although a potential effect remained a distinct possibility. Accordingly, we sought to answer this question by testing IL-10 in a shock model related to the one that had been successfully used to demonstrate its effect on LPS-induced lethal shock. In this model, the superantigen staphylococcal enterotoxin B (SEB) was used to induce widespread T-cell activation in vivo.[10] Superantigens are molecules that are recognized by all the T cells bearing a particular Vβ chain in their T-cell receptor (TCR).[11] In the case of SEB, all T cells using Vβ8 (i.e., Vβ8.1, 8.2, or 8.3) will be activated by this superantigen.[12] Many of these T cells will produce large amounts of some cytokines in response to the superantigen. As in the case of LPS, TNF-α has been implicated as one of the cytokines

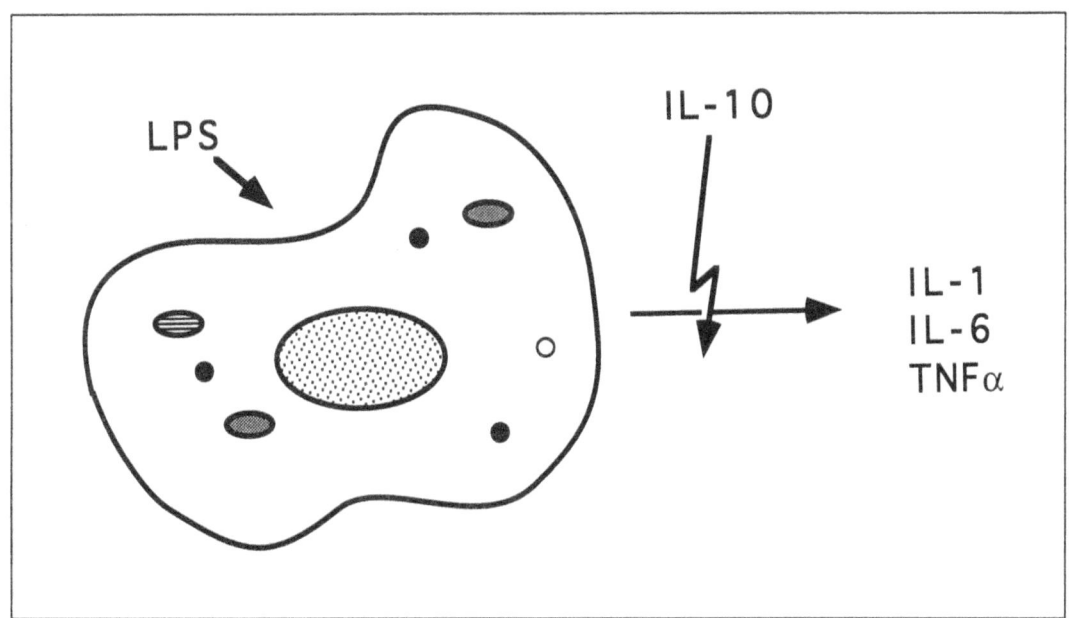

Fig. 13.2. IL-10 inhibits LPS-induced cytokine production in macrophages.

responsible for the toxic shock that occurs in the animals.[10] However, in sharp contrast to the LPS model where the target is the macrophage (which is likely to be the main producer of TNF-α), superantigen-induced lethal shock depends on cytokine production by T cells.[10] This was elegantly demonstrated by showing that SCID mice (deficient on T cells but not macrophages) are susceptible to LPS-induced, but not SEB-induced shock.[10] Superantigens, however, still require class II MHC positive antigen-presenting cells to activate T cells (Fig. 13.3).[13] This means that IL-10 could affect this antigen-presenting cell and inhibit its ability to activate T cells. In fact, in one of the original studies of the mechanism of action of IL-10,[5] we had already observed that IL-10 was able to inhibit the ability of macrophages to activate T cells in the presence of SEB.

To test the potential effect of IL-10 in this model, we pretreated mice with D-galactosamine (since using SEB to induce lethal shock works best when used in combination with D-galactosamine) and then administered SEB. A dose of 25 μg

of SEB/mouse (weighing 20-25 g/mouse) induced significant death (usually more than 80%) within 24 hours. However, if mice were pretreated with 10 μg of IL-10 the lethal effect of SEB was abrogated. In fact, most mice treated with IL-10 often showed no visible signs of distress. IL-10 could be given up to 30 minutes after SEB challenge and still induced significant protection.

This result indicates that IL-10 is able to inhibit T-cell activation in vivo. This is important because it validates many of the in vitro observations described in earlier studies. The likely mechanism of action is through its ability to inhibit the ability of macrophages to activate T cells. If this is correct, it also implies that IL-10 acts rapidly to abrogate the ability of resident macrophages to activate T cells triggered by a superantigen.

CONCLUDING REMARKS

The ability of IL-10 to induce protection for both the LPS and SEB models of lethal shock strongly suggests that it has both anti-inflammatory and immunomod-

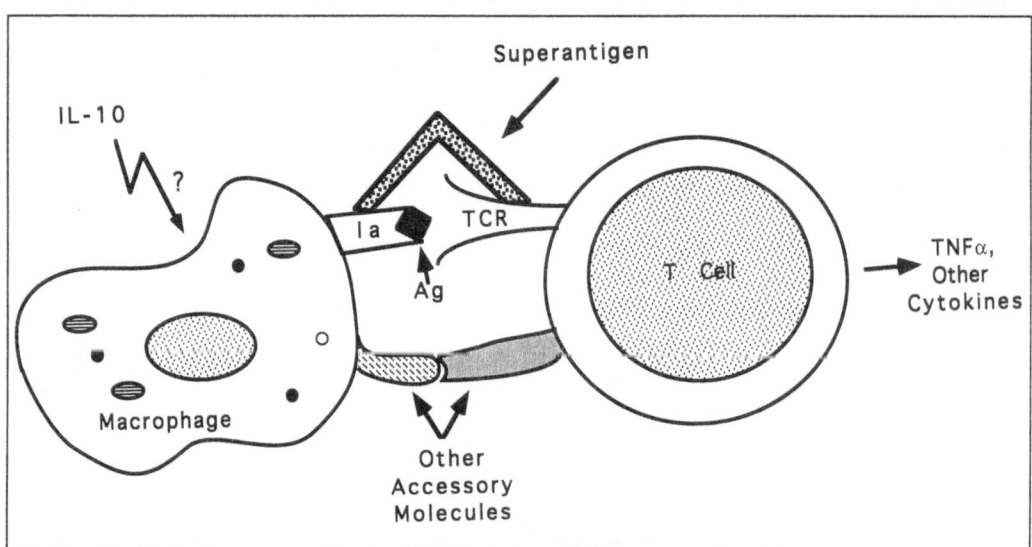

Fig. 13.3. Antigen-presenting cells process antigen (Ag) and present it to T cells in the context of class II MHC molecules (Ia). In this interaction, secondary accessory molecules are also important (see text). This results in cytokine production by T cells. Superantigens are molecules that interact with the constant portion of Ia molecules and the Vβ chain of the T-cell receptor, therefore activating many T cells bearing particular Vβ chains.

ulatory effects in vivo. This conclusion makes IL-10 an excellent candidate for a physiological anti-inflammatory as well as a mediator involved in the downregulation of immune responses. IL-10 is likely to be a specific anti-inflammatory agent, since it also induces the expression of other molecules with anti-inflammatory properties, like the IL-1 receptor antagonist.[14] Several other chapters in these reviews have reached the same conclusion. However, the lethal shock models provide dramatic evidence for the effects of IL-10, since the outcome could not be more clearcut than life or death of the experimental animals. Finally, IL-10 is likely to be effective in both gram-negative (LPS-mediated) and gram-positive (superantigen-mediated) sepsis, implying a broader potential for protection.

These results have potential clinical implications for some human diseases. A classic case is toxic shock syndrome, where another superantigen (toxic shock toxin) activates large numbers of T cells in vivo with potentially life-threatening clinical consequences.[15] Another hypothesis, favored by certain immunologists, envisions that autoimmune diseases may result from exposure to certain superantigens.[12,16] Following this exposure, a large number of T cells may become activated in vivo, and may not return to a "normal" resting state. This may result in increased avidity for some endogenous antigens that could be attacked by some of these cells. If this is the case, IL-10 could have potential as a prophylactic agent in the prevention of autoimmune diseases following exposure to superantigens.

Other cytokines (namely IL-4 and TGF-β) have been described to have potential anti-inflammatory and immunomodulatory properties. Yet, only IL-10 has demonstrated surprising effectiveness in vivo by itself. Its potential use in combination therapy is another possibility. At this point, there is reason to be optimistic about the potential use of IL-10 in a number of inflammatory and immune disorders.

REFERENCES

1. Zlotnik A, Moore K. Interleukin-10. Cytokine 1991; 3:366-71.
2. Thompson-Snipes L, Dhar V, Bond M et al. Interleukin-10: A novel stimulatory factor for mast cells and their precursors. J Exp Med 1991; 173:507-10.
3. MacNeil I, Suda T, Moore K et al. IL-10: A novel cytokine growth cofactor for mature and immature T cells. J Immunol 1990; 145:4167-73.
4. Go N, Castle B, Barrett R et al. Interleukin-10 (IL-10), a novel B cell stimulatory factor: Unresponsiveness of X chromosome-linked immunodeficiency B cells. J Exp Med 1990; 172:1625-31.
5. Fiorentino D, Zlotnik A, Vieira P et al. IL-10 acts on the antigen-presenting cell to inhibit cytokine production by Th1 cells. J Immunol 1991; 146:3444-51.
6. Fiorentino D, Zlotnik A, Mosmann T et al. IL-10 inhibits cytokine production by activated macrophages. J Immunol 1991; 147:3815-22.
7. Howard M, Muchamuel T, Andrade S et al. Interleukin-10 protects mice from lethal endotoxemia. J Exp Med 1993; 177:1205-08.
8. Gerard C, Bruyns C, Marchant A et al. Interleukin-10 reduces the release of tumor necrosis factor and prevents lethality in experimental endotoxemia. J Exp Med 1993; 177:547-50.
9. Galanos C, Freudenberg M, Reutter W. Galactosamine-induced sensitization to the lethal effects of endotoxin. Proc Natl Acad Sci 1979; 76:5939-43.
10. Miethke T, Wahl C, Heeg K et al. T-cell mediated lethal shock triggered in mice by the superantigen staphylococcal enterotoxin B: Critical role of tumor necrosis factor. J Exp Med 1992; 175:91-8.
11. Drake C, Kotzin B. Superantigens: biology, immunology and potential role in disease. J Clin Immunol 1992; 12:149-61.
12. Marrack P, Blackman M, Kushnir E et al. The toxicity of staphylococcal enterotoxin b in mice is mediated by T cells. J Exp Med 1990; 171:455-64.

13. Scholl PR, Diez A, Geha RS. Staphylococcal enterotoxin B and toxic shock syndrome toxin-1 bind to distinct sites on HLA-DR and HLA-DQ molecules. J Immunol 1989; 143:2583-88.

14. Cassatella M, Meda L, Gasperini S et al. Interleukin-10 upregulates IL-1 receptor antagonist production from lipopolysaccharide-stimulated leukocytes by delaying mRNA degradation. J Exp Med 1994; 179:1695-99.

15. Todd JK. Toxic shock syndrome. Clin Microbiol Rev 1988; 1:432-46.

16. Misfeldt ML. Microbial "superantigens". Infect Immun 1990; 58:2409-13.

CHAPTER 14

INHIBITION OF CELL-MEDIATED IMMUNE RESPONSES IN VIVO BY RECOMBINANT IL-10

Fiona Powrie and Robert L. Coffman

INTRODUCTION

One of the major advances in cellular immunology in the last decade has been the concept that distinct subsets of CD4[+] T cells regulate cell-mediated and humoral immune responses. Functional specialization of the CD4[+] T-cell population was originally proposed based on the finding that mouse CD4[+] T-cell clones could be subdivided into two major subtypes based on their cytokine repertoire; Th1 cells produce, amongst other cytokines, IFN-γ, TNF-α and β, and IL-2 and are efficient activators of macrophages, whereas Th2 cells are the principal stimulators of antibody production and produce IL-4, IL-5, IL-6 and IL-10, but not IFN-γ or IL-2.[1] Differential activation of Th1 and Th2 cells therefore provides a cellular basis for the long standing observation that cell-mediated and humoral immune responses are often mutually exclusive.[2]

The reciprocal nature of Th1 and Th2 responses prompted the search for molecules produced by one of the subsets which could antagonize the functions of the other. This approach led to the identification of murine and then human IL-10.[3,4] Originally defined as a product of Th2 cells that inhibited induction of cytokine synthesis and proliferation by Th1 clones,[5] IL-10 has been extensively studied in recent years.[6] Numerous studies have documented the inhibitory effects of IL-10 on Th-1 cytokine synthesis and macrophage activation in vitro.[7-10] However until recently little information was available on the functions of IL-10 in vivo. In the first part of this chapter we summarize our experiments to determine the effects of IL-10 on regulation of Th1 responses to a protozoan parasite in

Interleukin-10, edited by Jan E. deVries and René de Waal Malefyt.

vivo and compare this with the activities of IL-4. In the latter part we discuss the role of IL-10 as a potential therapeutic for the inhibition of pathogenic inflammatory responses in vivo, focusing on a murine model of inflammatory bowel disease as an example.

PRODUCTION OF IL-4 AND IL-10 IN RESPONSE TO *L. MAJOR* ANTIGENS IN VITRO INHIBITS IFN-γ PRODUCTION

Lack of cell-mediated immunity is the hallmark of a number of chronic infectious diseases such as lepromatous leprosy, filiriasis and visceral leishmaniasis and it has been suggested that cytokines produced by Th2 cells may in part account for this.[11] Most studies have focused on the role of IL-10 in this process. Thus, neutralization of IL-10 in cultures of spleen cells from mice infected with *Schistosoma mansoni* enhanced IFN-γ production in response to parasite antigens.[12] However more recently it has been appreciated that IL-4 can also inhibit IFN-γ production by CD4+ T cells[13-17] although unlike IL-10, IL-4 was not active in the original cytokine synthesis inhibition assay with Th1 clones as targets.[5]

In order to study the regulatory effects of both IL-4 and IL-10 on normal Th1 responses, we have used a murine model of leishmaniasis. This is a particularly useful infectious disease model as polarized Th1 or Th2 responses to the parasite develop depending on the genetic background of the mice. Most strains of mice mount an IFN-γ dependent Th1 response to the parasite and heal the lesion. In contrast BALB/c mice mount a Th2 response to the parasite, characterized by production of high IL-4 and low IFN-γ, which is ineffective at controlling the growth of the parasite, which disseminates to the viscera with fatal consequences.[18] Abrogation of the Th2 response in BALB/c mice by administration of anti-IL-4 mAbs within the first 2 weeks of infection results in a healing pattern of disease, indicating that the Th2 response is not only ineffective at controlling the infection but actively prevents the development of protective Th1 immunity.[19,20]

To determine whether IL-4 or IL-10 production accounted for the lack of IFN-γ production by lymph node cells (LNC) isolated from BALB/c mice 1 month after *L. major* infection, LNC were stimulated with *L. major* antigen together with anti-IL-4 or anti-IL-10 mAbs. Neutralization of either IL-4 or IL-10 led to 3-5-fold elevations in IFN-γ production, demonstrating that in vitro at least concomitant production of both IL-4 and IL-10 inhibited expression of Th1 function.[21] Similar results were obtained with purified CD4+ T cells and naive T-cell depleted splenocytes as APC.[22] Recently we have shown that during the dominant Th2 response to *L. major* the CD4+ T cells which produce IFN-γ can be distinguished from the CD4+ T cells that produce IL-4 and IL-10 based on the level of expression of the CD45RB antigen.[22] These results suggest that Th1 and Th2 cells co-exist in mice with chronic leishmaniasis, but that Th1 effector function is inhibited, at least in part by the dominant IL-4 and IL-10 producing Th2 cells. These studies may be relevant to human visceral leishmaniasis as neutralization of IL-10 and to a lesser extent IL-4 led to enhanced antigen specific proliferation and IFN-γ production from the PBL of patients with visceral leishmaniasis.[23]

When CD4+ T cells were isolated from mice mounting a highly polarized Th1 response to *L. major* (anti-CD4 pre-treated BALB/c mice which had healed their *L. major* infection) addition of recombinant IL-10 (rIL-10) or rIL-4 led to a maximal 50% inhibition of antigen induced IFN-γ synthesis whereas 75% inhibition was obtained if IL-4 and IL-10 were added together indicating an additive effect of these two cytokines.[21]

IL-4 AND IL-10 SYNERGIZE TO INHIBIT SECONDARY TH1 RESPONSES IN VIVO

Given the additive inhibition of Th1 responses in vitro with a combination of IL-4 and IL-10, we tested the effects of systemic administration of IL-4 and/or IL-10 on the development of delayed-type hypersensitivity (DTH) responses to *L. major* antigens in mice which had healed their *L. major* infection. As expected these mice mounted a significant DTH reaction when injected with soluble *L. major* antigens. Footpad swelling peaked between 48h and 72h indicating a classical tuberculin-type DTH. In some mice rIL-4, rIL-10 or a combination of both were administered intraperitoneally starting 12h prior to antigen challenge and continuing at 12 hourly intervals up to 48h after antigen challenge. At 24h after antigen challenge footpad swelling was significantly inhibited (60%) in mice that received a combination of IL-4 and IL-10, whereas there was little effect of administration of either cytokine alone. By 48h high doses (20 μg/injection) of IL-4 or IL-10 were able to inhibit footpad swelling however five-fold lower doses each of a combination of IL-4 and IL-10 inhibited footpad swelling by 70%.[21] Histological analysis of the footpads showed that treatment with IL-4 and IL-10 not only inhibited footpad edema but significantly inhibited leukocytic infiltration into the site of antigenic challenge. Somewhat suprisingly, analysis of antigen induced cytokine production in vitro by lymph node cells draining the site of antigenic challenge revealed that IFN-γ synthesis was significantly inhibited (70%), compared to control mice, in groups that received treatment with IL-4 or IL-10 alone, as well as those that received a combination of both.[21] Since LNC were removed three days after the last dose of cytokine, the effects of IL-4 and IL-10 on IFN-γ synthesis were long-lived compared with the short half lives of these molecules in vivo. Administration of IL-4 or IL-10 had not caused a general T-cell unresponsiveness as IL-3 responses to *L. major* antigen were unaffected. There was no evidence that these treatments which inhibited IFN-γ production led to an outgrowth of Th2 cells as IL-4 production was undetectable, suggesting IL-4 and IL-10 acted to inhibit the effector arm of the immune response. Taken together these results showed that both IL-4 and IL-10 could inhibit Th1 responses in vivo, but

that optimal inhibition of a complex readout of Th1 activity, such as DTH required a combination of both cytokines.[21]

Although DTH responses correlate with Th1 immunity and Th1 but not Th2 clones were able to adoptively transfer DTH responses to naive recipients,[24] the cytokines involved in the process are not well defined. IFN-γ has been shown to stimulate lymphocyte recruitment into DTH sites in the rat[25] and neutralization of IFN-γ in vivo partially inhibited the adoptive transfer of DTH responses with Th1 clones in vivo.[26] TNF has been shown to be a critical component of contact hypersensitivity responses, another form of DTH.[27] As IL-4 and IL-10 treatment inhibited IFN-γ production by LN cells draining the site of antigen challenge and both molecules have been shown to inhibit TNF-α production by activated macrophages[9] it seemed likely that IL-4 and IL-10 inhibited the DTH through a reduction in IFN-γ and TNF levels. However, treatment of mice with neutralizing mAbs to IFN-γ and/or TNF failed to inhibit the DTH and mimic the effects of administration of IL-4 and IL-10 suggesting another mechanism.

INHIBITION OF INFLAMMATORY BOWEL DISEASE IN MICE BY ADMINISTRATION OF RIL-10

Th1 responses are frequently beneficial, especially for the control of intracellular pathogens, however under some circumstances these host protective responses can induce immune pathology, as occurs in cerebral malaria[28] and tuberculoid leprosy.[29] The finding that systemic administration of IL-4 and IL-10 together effectively inhibited secondary Th1 effector functions in vivo suggested that this combination of cytokines may be useful for the treatment of chronic Th1 mediated inflammatory diseases.

We have recently described a model of inflammatory bowel disease (IBD) that developed spontaneously in C. B-17 *scid* mice after transfer of a subset of CD4+ T cells (identified by high level expression of the CD45RB antigen) from BALB/c mice. The reciprocal CD45RB^{low} CD4$^+$ T-cell subset not only did not induce colitis but, if co-transferred with the CD45RB^{high} population, prevented induction of colitis.[30] The colitis, which involved the cecum, colon and rectum, was characterized by a severe transmural lymphocytic infiltrate, involving the lamina propria, serosa and muscularis. There was also extensive epithelial cell hyperplasia and destruction of mucin secreting cells. Interestingly inflammation was predominantly restricted to the large intestine, the small intestine was generally not involved, and a small percentage of mice had evidence of gastritis, thyroiditis, hepatitis and lung inflammation.[30] These data provide evidence that normal BALB/c mice have CD4$^+$ T cells capable of inducing a chronic inflammatory response when transferred to immunodeficient recipients and that under normal circumstances this population is held in check by a phenotypically distinguishable subpopulation of CD4$^+$ T cells.

CD4$^+$ T cells isolated from the lamina propria of the colon in mice with colitis produced high levels of IFN-γ, but low levels of IL-4 and IL-10 upon polyclonal stimulation in vitro (F. Powrie et al, manuscript submitted). Analysis of cyto-kine mRNA from the colon of mice with colitis revealed high levels of IFN-γ and TNF-α mRNA, but low levels of IL-4 and IL-10 mRNA, similar to those in mice protected from colitis by transfer of both CD45RB^{high} and CD45RB^{low} subpopulations.[30] These data suggested that the colitis was due to the expansion of Th1 cells. To test this directly, mice reconstituted with CD45RB^{high} CD4$^+$ T cells were treated in the first two-weeks after T cell reconstitution with anti-IFN-γ mAbs. This treatment significantly inhibited the development of colitis as assessed 12 weeks after T-cell reconstitution. Anti-TNF also inhibited development of severe disease, however in contrast to anti-IFN-γ, anti-TNF had to be administered weekly to prevent disease (F. Powrie et al, manuscript submitted). Further, disease developed when anti-TNF treatment was stopped, suggesting anti-TNF worked by inhibiting effector function

rather than inducing a more permanent change in the immune response.

The finding that colitis induced by CD45RB^{high} CD4$^+$ T cells was prevented by anti-IFN-γ or anti-TNF provided good evidence for an IFN-γ and TNF dependent Th1 mediated pathogenesis. As administration of rIL-10 significantly inhibited the development of secondary Th1 responses to *L. major* we tested whether IL-10 could effect the development of the chronic Th1 inflammatory response in the colon. Mice restored with CD45RB^{high} CD4$^+$ T cells were treated daily with subcutaneous injections of PBS or PBS containing 20μg rIL-10. Mice sacrificed after 8 weeks of treatment with IL-10 were significantly protected from colitis (1/19 developed severe disease) compared with control, PBS treated mice (12/15 developed severe disease). Treatment of mice for 8w with rIL-10 did not appear to have caused a long lasting modulation of the immune response as colitis developed in mice 4w after the last treatment with IL-10 (F. Powrie et al, manuscript submitted). Prevention of colitis by the administration of IL-10 suggests that IL-10 may be important in the normal regulation of intestinal immune responses. Consistent with this idea, mice which have had their IL-10 gene inactivated by homologous recombination developed a severe enterocolitis.[31] It remains to be established whether IL-10 is involved in the inhibition of disease by the CD45RB^{low} population.

Mice treated with rIL-10 had lower levels of IFN-γ and TNF-α mRNA in the colon, compared to mice with colitis, suggesting IL-10 may prevent development of colitis via inhibition of IFN-γ and TNF-α, which were shown to be important mediators in the disease. The fact that protection with IL-10 was transitory and dependent on the continued presence of exogenous IL-10 suggested that IL-10 acted to prevent activation of the inflammatory response rather than modulating Th subsets. It may be that treatment with IL-4 and IL-10 together may induce a more permanent inhibition of colitis and studies are currently underway to test this.

While the inflammatory bowel disease induced in *scid* mice is not identical to IBD that occurs in human, there are some clear similarities, particularly with Chrons' disease which in contrast to ulcerative colitis is characterized by a transmural infiltrate with a high proportion of activated T cells[32] which produce IL-2 and IFN-γ.[33] The finding that IL-10 prevented development of colitis in this model suggests that IL-10 may be a useful therapeutic for the treatment of IBD in man.

Systemic administration of IL-10 in vivo was shown to protect mice from LPS[34] and SEA induced lethal endotoxemia.[35] Recently, treatment of non-obese diabetic (NOD) mice with rIL-10 was shown to significantly delay the onset and reduce the incidence of diabetes. Interestingly mice treated with IL-10 for 15-17 weeks were significantly protected from diabetes for up to 7 months, suggesting IL-10 mediated a more long-term effect on the immune response to the islets.[36] However, the effects of IL-10 may not be so simple, recently it has been shown by Sarvetnick and colleagues that expression of IL-10 under the rat insulin promoter led to leukocytic infiltration but no destruction of the insulin producing β cells, revealing a hitherto unappreciated property of IL-10 as a leukocyte attractant.[37] Somewhat unexpectedly, expression of the transgene on a NOD background led to an accelerated onset and increased incidence of diabetes.[38] These results are in apparent conflict with those obtained with systemic administration of IL-10, suggesting differences in the efficacy of IL-10 to prevent inflammatory responses depending on the mode of delivery. For efficient use of IL-10 as a therapeutic in vivo further studies are required to determine the effects of high local verus systemic levels of IL-10 on a number of inflammatory responses.

ACKNOWLEDGEMENTS

DNAX Research Institute for Molecular and Cellular Biology is supported by Schering Plough Corporation.

REFERENCES

1. Mosmann TR, Coffman RL. TH1 and TH2 cells: different patterns of lymphokine secretion lead to different functional properties. Annu Rev Immunol 1989; 7:145-73.
2. Parish CR. The relationship between humoral and cell-mediated immunity. Transplant Rev 1972; 13:35-46.
3. Moore KW, Vieira P, Fiorentino DF et al. Homology of cytokine synthesis inhibitory factor (IL-10) to the Epstein-Barr virus gene BCRFI. Science 1990; 248:1230-34.
4. Vieira P, de Waal Malefyt R, Dang MN et al. Isolation and expression of human cytokine synthesis inhibitory factor cDNA clones: homology to Epstein-Barr virus open reading frame BCRFI. Proc Natl Acad Sci USA 1991; 88:1172-6.
5. Fiorentino DF, Bond MW, Mosmann TR. Two types of mouse T helper cell. IV. Th2 clones secrete a factor that inhibits cytokine production by Th1 clones. J Exp Med 1989; 170:2081-95.
6. Moore KW, O'Garra A, de Waal Malefyt R et al. Interleukin-10. Annu Rev Immunol 1993; 11:165-90.
7. Bogdan C, Vodovotz Y, Nathan C, Macrophage deactivation by interleukin 10. J Exp Med 1991; 174:1549-55.
8. de Waal Malefyt R, Abrams J, Bennett B et al. Interleukin 10(IL-10) inhibits cytokine synthesis by human monocytes: an autoregulatory role of IL-10 produced by monocytes. J Exp Med 1991; 174:1209-20.
9. Fiorentino DF, Zlotnik A, Mosmann TR et al. IL-10 inhibits cytokine production by activated macrophages. J Immunol 1991; 147:3815-22.
10. Fiorentino DF, Zlotnik A, Vieira P et al. IL-10 acts on the antigen-presenting cell to inhibit cytokine production by Th1 cells. J Immunol 1991; 146:3444-51.
11. Sher A, Gazzinelli RT, Oswald IP et al. Role of T-cell derived cytokines in the downregulation of immune responses in parasitic and retroviral infection. Immunol Rev 1992; 127:183-204.
12. Sher A, Fiorentino D, Caspar P et al. Production of IL-10 by CD4+ T lymphocytes correlates with down-regulation of Th1

cytokine synthesis in helminth infection. J Immunol 1991; 147:2713-6.

13. Peleman R, Wu J, Fargeas C et al. Recombinant interleukin 4 suppresses the production of interferon gamma by human mononuclear cells. J Exp Med 1989; 170:1751-6.

14. Vercelli D, Jabara HH, Lauener RP et al. IL-4 inhibits the synthesis of IFN-gamma and induces the synthesis of IgE in human mixed lymphocyte cultures. J Immunol 1990; 144:570-73.

15. Tanaka T, Hu LJ, Seder RA et al. Interleukin-4 suppresses interleukin 2 and interferon gamma production by naive T cells stimulated by accessory cell-dependent receptor engagement. Proc Natl Acad Sci USA 1993; 90:5914-18.

16. Powrie F, Menon S, Coffman RL. Interleukin-4 and interleukin-10 synergize to inhibit cell-mediated immunity in vivo. Eur J Immunol 1993; 23:2223-29.

17. Salgame P, Yamamura M, Bloom BR et al. Evidence for functional subsets of CD4$^+$ and CD8$^+$ T cells in human disease: lymphokine patterns in leprosy. Chem Immunol 1992; 54:44-59.

18. Locksley RM, Scott P. Helper T-cell subsets in mouse leishmaniasis: induction, expansion and effector function. Immunol Today 1991;12:(3) pA58-61.

19. Sadick MD, Heinzel FP, Holaday BJ et al. Cure of murine leishmaniasis with anti-interleukin 4 monoclonal antibody. Evidence for a T cell-dependent, interferon gamma-independent mechanism. J Exp Med 1990; 171:115-27.

20. Chatelain R, Varkila K, Coffman RL. IL-4 induces a Th2 response in Leishmania major-infected mice. J Immunol 1992; 148:1182-87.

21. Powrie F Coffman RL. Inhibition of cell-mediated immunity by IL4 and IL10. Res Immunol 1993; 144:639-43.

22. Powrie F, Correa-Oliveira R, Mauze S et al. Regulatory interactions between CD45RB^{high} and CD45RB^{low} CD4$^+$ T cells are important for the balance between protective and pathogenic cell-mediated immunity. J Exp Med 1994; 179:589-600.

23. Carvalho EM, Bacellar O, Brownell C et al. Restoration of IFN-γ production and lymphocyte proliferation in visceral leishmaniasis. 1994; 152:5949-56.

24. Cher DJ, Mosmann TR. Two types of murine helper T cell clone. II. Delayed-type hypersensitivity is mediated by TH1 clones. J Immunol 1987; 138:3688-94.

25. Issekutz TB, Stoltz JM, van der Meide P. Lymphocyte recruitment in delayed-type hypersensitivity. The role of IFN-gamma. J Immunol 1988; 140:2989-93.

26. Fong TA, Mosmann TR. The role of IFN-gamma in delayed-type hypersensitivity mediated by Th1 clones. J Immunol 1989; 143:2887-93.

27. Piguet PF, Grau GE, Hauser C et al. Tumor necrosis factor is a critical mediator in hapten induced irritant and contact hypersensitivity reactions. J Exp Med 1991; 173:673-79.

28. Grau GE, Piguet PF, Vassalli P et al. Tumor-necrosis factor and other cytokines in cerebral malaria: experimental and clinical data. Immunol Rev 1989;112:49-70.

29. Yamamura M, Wang XH, Ohmen JD et al. Cytokine patterns of immunologically mediated tissue damage. J Immunol 1992;149:1470-75.

30. Powrie F, Leach MW, Mauze S et al. Phenotypically distinct subsets of CD4$^+$ T cells induce or protect from chronic intestinal inflammation in C. B-17 *scid* mice. Int Immunol 1993; 5:1461-71.

31. Kuhn R, Lohler J, Rennick D, et al. Interleukin-10 deficient mice develop chronic enterocolitis. Cell 1993; 75:263-74.

32. Choy MY, Walker-Smith JA, Williams CB et al. Differential expression of CD25 (interleukin-2 receptor) on lamina propria T cells and macrophages in the intestinal lesions in Crohn's disease and ulcerative colitis. Gut 1990; 31:1365-70.

33. Breese E, Braegger CP, Corrigan CJ et al. Interleukin-2- and interferon-gamma-secreting T cells in normal and diseased human intestinal mucosa. Immunology 1993; 78:127-31.

34. Howard M, Muchamuel T, Andrade S et al. Interleukin 10 protects mice from lethal endotoxemia. J Exp Med 1993; 177:1205-08.

35. Bean AG, Freiberg RA, Andrade S et al.

QUESTIONNAIRE

Receive a FREE BOOK of your choice

Please help us out—Just answer the questions below, then select the book of your choice from the list on the back and return this card.

R.G. Landes Company publishes five book series: *Medical Intelligence Unit, Molecular Biology Intelligence Unit, Neuroscience Intelligence Unit, Tissue Engineering Intelligence Unit* and *Biotechnology Intelligence Unit.* We also publish comprehensive, shorter than book-length reports on well-circumscribed topics in molecular biology and medicine. The authors of our books and reports are acknowledged leaders in their fields and the topics are unique. Almost without exception, there are no other comprehensive publications on these topics.

Our goal is to publish material in important and rapidly changing areas of bioscience for sophisticated scientists. To achieve this goal, we have accelerated our publishing program to conform to the fast pace in which information grows in bioscience. Most of our books and reports are published within 90 to 120 days of receipt of the manuscript.

Please circle your response to the questions below.

1. We would like to sell our *books* to scientists and students at a deep discount. But we can only do this as part of a prepaid subscription program. The retail price range for our books is $59-$99. Would you pay $196 to select four *books* per year from any of our Intelligence Units–$49 per book–as part of a prepaid program?

    Yes   No

2. We would like to sell our *reports* to scientists and students at a deep discount. But we can only do this as part of a prepaid subscription program. The retail price range for our reports is $39-$59. Would you pay $145 to select five *reports* per year–$29 per report–as part of a prepaid program?

    Yes   No

3. Would you pay $39–the retail price range of our books is $59-$99–to receive any single book in our Intelligence Units if it is spiral bound, but in every other way identical to the more expensive hardcover version?

    Yes   No

To receive your free book, please fill out the shipping information below, select your free book choice from the list on the back of this survey and mail this card to:

R.G. Landes Company, 909 S. Pine Street, Georgetown, Texas 78626 U.S.A.

Your Name _____

Address _____

City _____ State/Province: _____

Country: _____ Postal Code: _____

My computer type is Macintosh_____ ; IBM-compatible _____ ; Other _____

Do you own ____ or plan to purchase ____ a CD-ROM drive?

AVAILABLE FREE TITLES

☐ Water Channels
Alan Verkman,
University of California-San Francisco

☐ The Na,K-ATPase:
Structure-Function Relationship
J.-D. Horisberger, University of Lausanne

☐ Intrathymic Development of T Cells
J. Nikolic-Zugic,
Memorial Sloan-Kettering Cancer Center

☐ Cyclic GMP
Thomas Lincoln, University of Alabama

☐ Primordial VRM System and the Evolution
of Vertebrate Immunity
John Stewart, Institut Pasteur-Paris

☐ Thyroid Hormone Regulation
of Gene Expression
Graham R. Williams, University of Birmingham

☐ Mechanisms of Immunological Self Tolerance
Guido Kroemer, CNRS Génétique Moléculaire et
Biologie du Développement-Villejuif

☐ The Costimulatory Pathway
for T Cell Responses
Yang Liu, New York University

☐ Molecular Genetics of Drosophila Oogenesis
Paul F. Lasko, McGill University

☐ Mechanism of Steroid Hormone Regulation
of Gene Transcription
M.-J. Tsai & Bert W. O'Malley, Baylor University

☐ Liver Gene Expression
François Tronche & Moshe Yaniv,
Institut Pasteur-Paris

☐ RNA Polymerase III Transcription
R.J. White, University of Cambridge

☐ src Family of Tyrosine Kinases in Leukocytes
Tomas Mustelin, La Jolla Institute

☐ MHC Antigens and NK Cells
Rafael Solana & Jose Peña,
University of Córdoba

☐ Kinetic Modeling of Gene Expression
James L. Hargrove, University of Georgia

☐ PCR and the Analysis of the T Cell Receptor
Repertoire
Jorge Oksenberg, Michael Panzara & Lawrence
Steinman, Stanford University

☐ Myointimal Hyperplasia
Philip Dobrin, Loyola University

☐ Transgenic Mice as an In Vivo Model
of Self-Reactivity
David Ferrick & Lisa DiMolfetto-Landon,
University of California-Davis and Pamela Ohashi,
Ontario Cancer Institute

☐ Cytogenetics of Bone and Soft Tissue Tumors
Avery A. Sandberg, Genetrix & Julia A. Bridge,
University of Nebraska

☐ The Th1-Th2 Paradigm and Transplantation
Robin Lowry, Emory University

☐ Phagocyte Production and Function Following
Thermal Injury
Verlyn Peterson & Daniel R. Ambruso,
University of Colorado

☐ Human T Lymphocyte Activation Deficiencies
José Regueiro, Carlos Rodríguez-Gallego
and Antonio Arnaiz-Villena,
Hospital 12 de Octubre-Madrid

☐ Monoclonal Antibody in Detection and
Treatment of Colon Cancer
Edward W. Martin, Jr., Ohio State University

☐ Enteric Physiology of the Transplanted Intestine
Michael Sarr & Nadey S. Hakim, Mayo Clinic

☐ Artificial Chordae in Mitral Valve Surgery
Claudio Zussa, S. Maria dei Battuti Hospital-Treviso

☐ Injury and Tumor Implantation
Satya Murthy & Edward Scanlon,
Northwestern University

☐ Support of the Acutely Failing Liver
A.A. Demetriou, Cedars-Sinai

☐ Reactive Metabolites of Oxygen and Nitrogen
in Biology and Medicine
Matthew Grisham, Louisiana State-Shreveport

☐ Biology of Lung Cancer
Adi Gazdar & Paul Carbone,
Southwestern Medical Center

☐ Quantitative Measurement
of Venous Incompetence
Paul S. van Bemmelen, Southern Illinois University
and John J. Bergan, Scripps Memorial Hospital

☐ Adhesion Molecules in Organ Transplants
Gustav Steinhoff, University of Kiel

☐ Purging in Bone Marrow Transplantation
Subhash C. Gulati,
Memorial Sloan-Kettering Cancer Center

☐ Trauma 2000: Strategies for the New Millennium
David J. Dries & Richard L. Gamelli,
Loyola University

Interleukin 10 protects mice against staphylococcal enterotoxin B-induced lethal shock. Infect Immun 1993; 61:4937-39.

36. Pennline KJ, Roque-Gaffney E, Monahan M. Recombinant human IL-10 prevents the onset of diabetes in the non-obese diabetic mouse. Clin Immunol and immunopathol 1994; 71:169-75.

37. Wogensen L, Huang X, Sarvetnick N. Leukocyte extravasation into the pancreatic tissue in transgenic mice expressing interleukin 10 in the islets of Langerhans. J Exp Med 1993; 178:175-85.

38. Wogensen L, Lee MS, Sarvetnick N. Production of interleukin 10 by islet cells accelerates immune-mediated destruction of beta cells in non-obese diabetic mice. J Exp Med 1994; 179:1379-84.

CHAPTER 15

IL-10 TRANSGENIC MICE

Myung-Shik Lee and Nora Sarvetnick

INTRODUCTION

A number of in vitro studies demonstrated strong immunoinhibition by IL-10, and suggest a role for it as a potential agent for the treatment of autoimmune diseases such as type I diabetes and multiple sclerosis, or for the prolongation of allograft survival. However, in vivo effects of IL-10 have not been addressed well, partly because of a lack of suitable animal models to test them.

We produced transgenic mice expressing IL-10 in the pancreatic β islets directed by the insulin promoter (Ins-IL-10 mice) because the pancreas is relatively easy to manipulate for the study of allograft rejection, and good autoimmune diabetes models are available for the study of its in vivo effects. We observed unexpected findings, which indicate that extrapolation of in vitro experiments to the in vivo situation is not always predictable because of complex interactions occurring between various limbs of the immune network.

PHENOTYPE OF INS-IL-10 TRANSGENIC MICE

Scientists characterized the phenotype of Ins-IL-10 mice before studying the immunological impact of the transgenic expression of IL-10. Surprisingly, lymphocyte infiltration in the periislet area was observed, which begins at 1 week of age and progresses to the surrounding exocrine area as the transgenic mice aged.[1] The infiltrating cells were mainly composed of macrophages at early stages of inflammation but at a later stage, CD4+ cells, CD8+ cells, and B lymphocytes dominated the infiltrating cells. However, islets remained intact for the entire observation period of more than 1 year, and transgenic mice never became diabetic for that period.[1] To investigate the mechanism of this unexpected periislet inflammation, morphological and functional changes of endothelium were studied. Endothelial cells appeared to be activated because they hyperexpressed ICAM-1 and MHC class II molecules. Furthermore, some endothelial cells were positively stained with MECA-367 and MECA-79

Interleukin-10, edited by Jan E. deVries and René de Waal Malefyt.
© 1995 R.G. Landes Company.

antibodies specific for a mucosal high endothelial venule (HEV) antigen and a peripheral lymph node HEV antigen, respectively.[2,3] These changes suggested differentiation of endothelium toward HEV that plays a critical role in the lymphocyte homing to the chronic inflammatory foci. The endothelial changes could contribute to or elicit extravasation of lymphocytes into the inflammatory areas by facilitating lymphocyte-endothelial interactions. These unexpected findings suggest that IL-10 could be a recruitment signal for lymphocyte migration in vivo, which may be mediated by IL-10 itself or may be secondary to other cytokines induced in response to IL-10.

FUNCTIONAL ASPECTS OF TRANSGENIC EXPRESSION OF IL-10

Although inflammatory changes were observed around the pancreatic islets, we thought those changes are not inconsistent with the immunosuppressive effects of IL-10 observed in vitro because islets remained intact and inflammatory cells were not destructive to islets. This assumption led us to study the functional effects of pancreatic IL-10 on the allograft rejections and autoimmune destruction of pancreatic islets.

First, we investigated possible attenuation of allograft rejection by transgenic production of IL-10 (Fig. 15.1A). Fetal pancreata were grafted under the kidney capsule of MHC-incompatible mice. Histological examination of the grafted tissue revealed complete rejection of the transgenic fetal pancreas 2 weeks after the allograft. Because the fetal pancreas is more immunogenic than other fetal tissue, we grafted adult transgenic islets to MHC-incompatible hosts. Here again, complete rejection of the transgenic islets was noted. There was no difference in the time frame of rejection between nontransgenic and transgenic islet or fetal pancreas grafts.[4] The failure to prevent allograft rejection by transgenic IL-10 expression was not due to low production of IL-10 because substan-

tial production of IL-10 was observed in the culture supernatant of transgenic islets.[4]

Next, we assessed the ability of IL-10 to protect pancreatic islets from destruction in a transgenic model where immune sensitization to islets occurs as a cross-reaction with viral antigens on islet cells after lymphocytic choriomeningitis virus (LCMV) infection (Fig. 15.1B).[5] Ins-IL-10 mice were crossed with RIP-LCMV-GP or RIP-LCMV-NP mice, and the litters were infected with LCMV to elicit (auto)immune destruction of the transgenic islets expressing LCMV antigens. Development of diabetes after LCMV infection of IL-10/LCMV-NP or IL-10/LCMV-GP double transgenic mice was accelerated rather than delayed compared to that in LCMV-NP or LCMV-GP single transgenic littermates, implying that local production of IL-10 does not inhibit immune-mediated destruction of pancreatic islets.[4] Histopathological studies also revealed intraislet infiltration of lymphocytes and destructive changes in the pancreatic islets of the double transgenic mice not ameliorated by local IL-10 production.

In another attempt to investigate the effect of transgenic IL-10 production on autoimmune processes, we tried to backcross Ins-IL-10 mice to NOD mice that develop autoimmune diabetes and to make NOD mice expressing IL-10 in the pancreatic islets (Fig. 15.1C). While backcrossing, to our surprise some of IL-10-transgenic NOD N2 mice became diabetic. Analysis of MHC type in N2 backcross revealed that over 90% of transgenic NOD N_2 with H-$2^{g7/g7}$ haplotype became diabetic before the age of 10 weeks. None of the nontransgenic NOD N2 mice with H-$2^{g7/g7}$ haplotype were diabetic. The percentage of islets showing insulitis with β cell necrosis was also very different between transgenic and nontransgenic NOD N_2 mice with H-$2^{g7/g7}$ haplotype (52% vs. 0%). IL-10-transgenic heterozygous NOD N_2 (H-$2^{g7/d}$) were not diabetic; however, 33% of IL-10-transgenic NOD N_3 with H-$2^{g7/d}$ became diabetic.[6] These data indicate that IL-10 accelerated insulitis and diabetes in

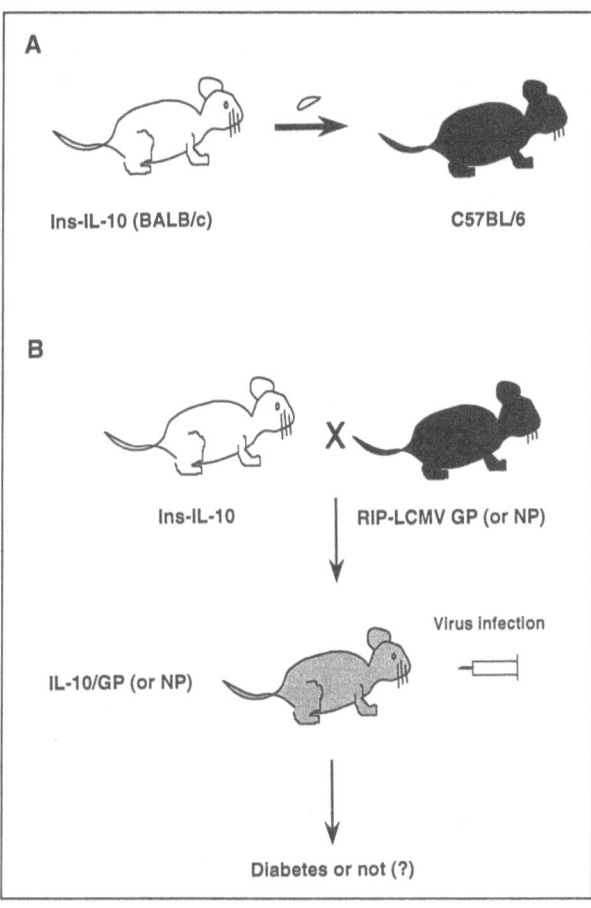

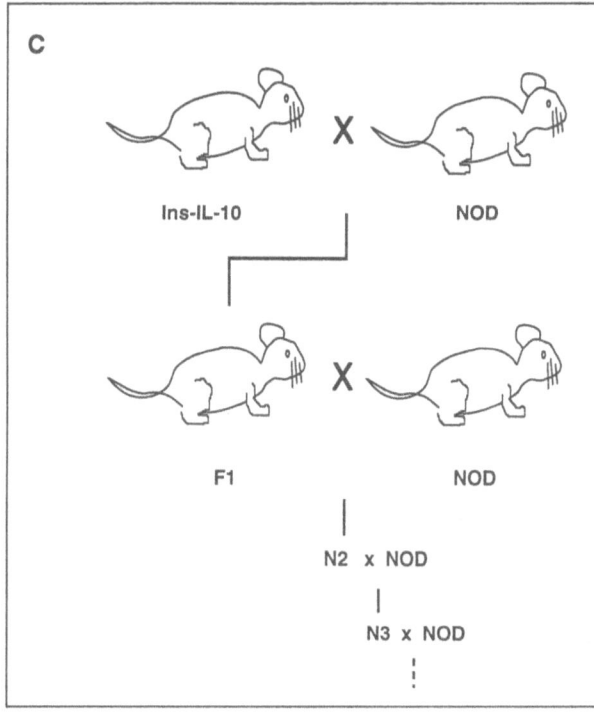

Fig. 15.1A & B. Experimental models to study the in vivo effects of IL-10. (A) IL-10-transgenic fetal pancreata or adult islets were allografted under the kidney capsule to address if IL-10 could inhibit allograft rejection. (B) Double transgenic mice expressing both pancreatic IL-10 and LCMV GP (or NP) antigen were produced and infected with LCMV to observe if IL-10 could inhibit LCMV-induced autoimmune diabetes. J. Clin Invest 1994; 93:1332.

Fig. 15.1C. Experimental model to study the in vivo effects of IL-10. (C) IL-10-transgenic mice were back-crossed to NOD mice to study the effect of pancreatic IL-10 on autoimmune diabetes.

NOD backcross. Furthermore, it is suggested that IL-10 might overcome genetic barriers imposed by non-MHC diabetes resistance genes or even by a single dose of non-H-2[87] MHC. The mechanism of these findings is not clear. The acceleration of diabetes is not likely to result from overproduction of a foreign protein by β islet cells because Moritani et al observed that transgenic NOD mice expressing IL-10 in the pancreatic α islet cells directed by glucagon promoter had also accelerated onset of diabetes in both sexes.[7] The transgenic NOD mice express IL-10 in α islet cells but not in β islet cells making the possibility of storage disease effect an unlikely explanation for the acceleration of diabetes in their model. One possibility is the disturbance of balance between Th1 and Th2 responses by IL-10 because IL-10 can inhibit Th1 responses.[8] This possibility is supported by the predominance of Th2-subset of CD4+ cells in the pancreas of transgenic NOD mice expressing IL-10 in the α islet cells and that of Ins-IL-10 mice.[6,7] An analysis of cytokine profiles in the pancreas of unmanipulated NOD mice also disclosed a preponderance of Th2-like cells with IL-4 and IL-10 expression among the intraislet CD4+ cells.[9]

CONCLUDING REMARKS

Our experiments to examine in vivo immunoinhibitory activity of IL-10 led to some findings that are contrary to widespread expectations. These results indicate that prediction of in vivo responses on the basis of in vitro findings is sometimes difficult, and underscore the complexity of immune system. However, it also should be kept in mind that our findings do not exclude the possibility of IL-10 as a systemic immunosuppressive agent against autoimmune diseases or graft rejection because systemic administration of IL-10 may have effects different from those of localized long-term production of IL-10. A recent report described reduction of diabetes incidence by systemic IL-10 administration.[10] The mechanism of the difference between our findings and those of systemic administration are currently being studied, and elucidation of true in vivo role of IL-10 awaits further elucidation.

REFERENCES

1. Wogensen L, Huang X, Sarvetnick N. Leukocyte extravasation into the pancreatic tissue in transgenic mice expressing interleukin 10 in the islets of Langerhans. J Exp Med 1993; 178:175-85.

2. Streeter PR, Berg EL, Rouse BTN et al. A tissue-specific endothelial cell molecule involved in lymphocyte homing. Nature 1988; 331:41-6.

3. Streeter PR, Rouse BTN, Butcher EC. Immunohistologic and functional characterization of a vascular addressin involved in lymphocyte homing into peripheral lymph nodes. J Cell Biol 1988; 107:1853-62.

4. Lee M-S, Wogensen L, Shizuru J et al. Pancreatic islet production of murine interleukin-10 does not inhibit immune-mediated tissue destruction. J Clin Invest 1994; 93:1332-38.

5. Oldstone MBA, Nerenberg M, Southern P et al. Virus infection triggers insulin-dependent diabetes mellitus in a transgenic model: role of anti-self (virus) immune response. Cell 1991; 65:319-31.

6. Wogensen L, Lee M-S, Sarvetnick N. Production of interleukin 10 by islet cells accelerates immune-mediated destruction of B cells in nonobese diabetic mice. J Exp Med 1994; 179:1379-84.

7. Moritani M, Yoshimoto K, Miyazaki J et al. Local production of IL-10 in pancreatic islet A-cells in transgenic NOD mice accelerates autoimmune insulitis and diabetes. (Abstract) Combined Meeting of The 8th International Lymphokine Workshop and The 4th International Workshop on Cytokines. Osaka, Japan, Oct 17-21, 1993.

8. Fiorentino DF, Zlotnik A, Vieira P et al. IL-10 acts on the antigen-presenting cell to inhibit cytokine production by Th1 cells. J Immunol 1991; 146:3444-51.

9. Anderson JT, Cornelius JG, Jarpe AJ et al. Insulin-dependent diabetes in the NOD mouse model II. β cell destruction in autoimmune diabetes in a Th2 and not a Th1 mediated event. Autoimmunity 1993; 14:113-22.

10. Penniline KJ, Roque-Gaffney E, Monohan M. Recombinant human IL-10 (rHUIL-10) prevents the onset of diabetes in the nonobese diabetic (NOD) mouse. Clin Immunol Immunopathol. In press.

=================CHAPTER 16=================

INTERLEUKIN-10 DEFICIENT MICE

Donna Rennick, Dan Berg, Ralf Kühn and Werner Müller

INTRODUCTION

After the mutation which inactivates the IL-4 gene had been introduced into embryonic stem cells, but before the IL-4-deficient mice[1] had been generated or analyzed, a new cytokine, designated interleukin-10 (IL-10) was cloned.[2] Many of the biological properties of this new cytokine such as the induction of increased expression of MHC class II molecules on B cells[3] and a strong inhibitory effect on the production of inflammatory cytokines by macrophages[4,5] were similar to those of IL-4. Consequently, it seemed logical to inactivate the IL-10 gene in the mouse germline to elucidate its main biological functions and in the future to combine the two mutations by crossing the mutant mice. As the entire genomic structure of the IL-10 gene was not known at the time of the gene inactivation construct, we decided to replace a portion of its putative first exon by the neomycin gene and to introduce a frame shift mutation into the coding region by destroying an EcoRI site of the genomic clone a site which was later found to be located in the third exon of the gene.[6,7]

Gene targeting has proven to be a very powerful method for the determination of gene function in vivo. The function of some genes may be relatively straight forward and follow the predictions deduced from previous work. The IL-4 deficient mouse mutant is, at least in part, such an example. IL-4 had been identified for a relatively long period of time[8,9] and its biological activities were well characterized[10] before the gene targeting was attempted. Treatment of mice with anti-IL-4 antibodies had suggested that IL-4 was essential for the production of antibodies of the IgE class and this point was proven by the IL-4 deficient mutant.[1] On the other hand, the strong effect of IL-4 on IgG1 antibody production was unexpected from studies using anti-IL-4 antibodies.[1]

Interleukin-10, edited by Jan E. deVries and René de Waal Malefyt.
© 1995 R.G. Landes Company.

The targeting of the IL-10 gene was accomplished at a time when the in vitro properties of IL-10 were not well characterized. Treatment of mice with anti-IL-10 antibodies resulted in the preferential loss of CD5+ B cells and antibody production associated with this B-cell subset.[11] This observation was not confirmed in the mutant.[7] Very much to our surprise, lymphocyte development and antibody responses were normal in the mutant mice, but most animals were growth retarded, anemic and suffered from chronic enter-ocolitis.[7] Therefore, IL-10 seems to be an essential immunoregulator of inflammatory processes and its absence leads to uncontrolled immune responses in the gut.

In this chapter we discuss how IL-10 function relates to the chronic enterocolitis in IL-10 deficient mice and which cells are involved in the inflammatory process. In addition, we will provide evidence that the anti-inflammatory property of IL-10 is not only essential in the gut, but that IL-10 deficient mice suffer from inducible uncontrolled inflammatory processes at other locations.

POTENTIAL MECHANISMS LEADING TO CHRONIC INFLAMMATORY BOWEL DISEASE IN IL-10 DEFICIENT MICE

After it was recognized that IL-10 deficient mice were afflicted with enterocolitis, we initiated studies to identify the underlying cause of their disease. The possibility that the mutants were infected with enteric pathogens was excluded by microbiological analyses at two diagnostic laboratories (Simonsen Laboratory, Gilroy, CA and Research Animal Diagnostic and Investigative Laboratory, Columbia, MO). In addition, IL-10 deficient mice, derived and housed under specific pathogen free (SPF) conditions, were found to develop disease albeit attenuated. Presently, we assume that the enterocolitis exhibited by IL-10 deficient mice results from the absence of the normal suppressive effect of IL-10 on immune-inflammatory responses to normal

enteric antigens. Results of in vitro assays suggest that IL-10 has several major immunosuppressive activities. It directly inhibits macrophage activation and production of numerous inflammatory cytokines.[4,5,12-14] Further, IL-10 indirectly suppresses natural killer (NK) cell function[15] and the generation of Th1 cells which are characterized by IL-2 and IFN-γ production.[13,16-19]

Considering these actions of IL-10, it seems likely that enterocolitis in IL-10 deficient mice is initiated and perpetuated by chronically activated macrophages, Th1 cells and NK cells. The uncontrolled release of cytokines by these immune-reactive cells could generate a series of events that amplify the inflammatory process. These include the activation of intestinal cells which synthesize additional pro-inflammatory mediators, the recruitment of inflammatory cells, and the induction of antibody synthesis. Accordingly, we found altered MHC class II expression by intestinal epithelium cells of IL-10 deficient mice and increased numbers of mononuclear cells, IgA- and IgG-producing plasma cells, neutrophils and eosinophils within focal gut lesions.[7] Given the array of cells that are involved and the types of biological mediators they produce, the intestinal lesions in IL-10 deficient mice may be caused by a number of mechanisms (e.g. antibody-dependent cell-mediated cytotoxicity, complement-mediated injury and cell damage mediated by arachidonic acid metabolites, proteases or reactive oxygen intermediates).

IMMUNOHISTOLOGICAL ANALYSES OF COLONIC LYMPHOCYTES

Recently, we have begun to analyze lymphocytes isolated from colons of affected mutants (Berg et al, unpublished data). The number of colonic lymphocytes recovered from the IL-10 deficient mice was only twice that recovered from normal animals. A two-fold difference was not unexpected because the highly infiltrated lesions in the mutants were intermittently dispersed among large amounts of normal tissue. When the composition of the lymphocyte subsets was analyzed by flow cytometry, the ratio of CD3$^+$ and B220$^+$ cells was the same in IL-10 deficient and control mice suggesting that in the mutants there was an equivalent increase in the absolute numbers of T and B cells. All T-cell subsets from IL-10 deficient mice were present in normal numbers with two exceptions. There was a 5-fold and 16-fold increase in the absolute numbers of CD4$^+$ αβTCR$^+$ cells and CD4$^+$CD8$^+$ αβTCR$^+$ cells, respectively. We assume that the frequency of these cells within focal lesions is much higher. In situ immunohistochemical analyses to prove this point are underway. We have also found that the majority of T cells isolated from IL-10 deficient mice, but only a minority of T cells isolated from normal mice, exhibit an activated/memory cell phenotype defined by high expression of Pgp-1 and reduced expression of CD45RB molecules. These results suggest, but do not prove, that activated CD4$^+$ T cells may be a causal factor in the development of enterocolitis in IL-10 deficient mice. Transplantation experiments are in progress to assess the ability of CD4$^+$ T cells from diseased mutants to initiate and sustain inflammatory bowel disease in immunodeficient recipients.

IL-10 DEFICIENT MICE EXHIBIT ABERRANT CYTOKINE PRODUCTION

Other studies have investigated the possibility that aberrant cytokine production in the gut of IL-10 deficient mice contributes to their disease. Immunohistochemical staining revealed high levels of TNF-α in frozen colonic sections. In addition to TNF-α, we detected spontaneous synthesis of IL-1 and IFN-γ in cultures of intestinal tissue from IL-10 deficient mice but not from control mice (Berg et al, unpublished results). IL-4 was undetectable in the same cell cultures. Macrophages are a major source of TNF-α and IL-1 whereas, Th1 cells and NK cells are a major source of IFN-γ. The increased production of such cytokines in the mucosa of mutant mice is

consistent with persistent activation of these three types of immune-reactive cells.

Analyses of other tissues in IL-10 deficient mice failed to detect spontaneous production of any cytokines suggesting that peripheral tissues are not undergoing a chronic inflammatory reaction. However, when macrophages from peripheral tissues of mutants were stimulated in culture with lipopolysaccharide (LPS), they produced 6- to 10-fold more IL-6 and TNF-α than LPS-stimulated cells from control mice. Moreover, the spleens of IL-10 deficient mice contained three-fold more IFN-γ producing CD4$^+$ T cells after activation with anti-CD3 antibodies.[7] IL-4 and IL-5, which are associated with Th2 cell responses, were undetectable. These results indicate that unregulated cytokine production by activated macrophages and Th1 cells may occur in any tissue of IL-10 deficient mice. However, under our specific housing conditions, it appears to be limited to the gut where persistent antigenic exposure is unavoidable.

We have not detected elevated levels of cytokines (i.e. IL-1, IL-6, TNF-α or IFN-γ) in the blood of IL-10 deficient mutants. Nevertheless, we assume that cytokines produced in the gut are released into the blood and affect the activities of other organs such as spleen, bone marrow and liver. This assumption is based on the systemic complications which occur in the mutants (i.e. splenomegaly, wasting, anemia, elevated platelet and white blood cell counts, increased sedimentation rates and high serum amyloid A levels).

In an effort to determine the contribution of various cytokines to the development of chronic enterocolitis, IL-10 deficient mice were infused with neutralizing antibodies specific for IL-6, TNF-α or IFN-γ. After 3 months of treatment with individual antibodies, there was no improvement in the disease exhibited by older animals and there was no decrease in the incidence of new disease in young animals (Berg et al, unpublished data). The results of our studies show that these cytokines by themselves do not play a major role in the initiation and/or maintenance of the inflammatory process. However, our results do not exclude the possibility that multiple cytokine interactions are involved. When diseased mutants were treated with IL-10 for 3 months, they showed dramatic improvement judged by their increased weight gain and survival rate. Furthermore, young mutants (3-4 weeks old) treated similarly with IL-10 remained healthy while their age-matched controls became anemic and exhibited significant inflammatory changes in their ceca and colons (Berg et al, unpublished data). IL-10 has been shown to inhibit the synthesis in vitro of a large array of cytokines by immune-activated cells. This broad suppressive effect of IL-10 may explain, in part, its successful use in the treatment of IL-10 deficient mice.

IL-10 DEFICIENT MICE EXHIBIT OTHER ABNORMAL IMMUNE-INFLAMMATORY REACTIONS

The finding that IL-10 deficient mice spontaneously develop enterocolitis demonstrates that IL-10 is an essential immunoregulator in the intestinal tract. However, it is known that many cell types can produce IL-10 and that they are widely distributed throughout the body.[5,20-27] This immediately raises questions about the importance of this cytokine in regulating immune reactions at other anatomical sites. Therefore, we investigated the immunological responses of IL-10 deficient mice to several antigens using different routes of administration.

The possibility that IL-10 functions as an immunoregulator of skin reactions can be inferred from the fact that keratinocytes produce IL-10 following exposure to skin-sensitizing compounds.[27] Two types of responses were elicited in IL-10 deficient mice by the epicutanous application of croton oil and oxazalone. Croton oil induces an immediate "irritant" reaction whereas oxazalone induces a CD8+ T cell-dependent contact hypersensitivity reaction[28] in individuals that have been previously sensitized. After the ears of IL-10 deficient and control mice were painted with croton oil or oxazalone, their immune-inflammatory responses were compared by measuring ear swelling. Both the magnitude and the duration of the inflammatory reactions elicited by these compounds were dramatically increased in IL-10 deficient mice (Berg et al, unpublished data).

Furthermore, in contrast to normal mice, the IL-10 mutants were extremely intolerant of high concentrations of croton oil. Their ears showed extensive necrosis and/or hemorrhage within 24 hours of exposure. Because these pathological changes have been shown to be mediated by TNF-α,[29,30] we conjectured that the uncontrolled production of TNF-α by the mutants led to this deleterious outcome. This assumption proved to be correct because tissue necrosis could be prevented by administering anti-TNF-α antibodies at the time of antigenic challenge. Based on these findings, we have concluded that IL-10 functions as a natural anti-inflammatory agent in the skin where it diminishes damaging side effects that accompany immune-inflammatory reactions.

Results of studies in vitro have shown that IL-10 inhibits the actions of CD4+ Th1 cells which are known to generate delayed type hypersensitivity (DTH) reactions through their release of IFN-γ.[31,32] Therefore, comparing the DTH responses of IL-10 deficient and normal mice was of particular interest. In our experiments, mice were primed by intraperitoneal injection of keyhole limpet hemocyanin (KLH) and challenged 7 days later with an intradermal injection of KLH in the footpad. The degree of footpad swelling has been shown to correlate directly with IFN-γ-dependent accumulation of activated immune-inflammatory cells.[32] We found that the footpads of IL-10 deficient mice were greatly increased in size relative to those of control mice (Berg et al, unpublished data). In addition, the chronic phase of the reaction was significantly prolonged and ultimately led to more tissue damage and scar formation. When spleen cells were harvested and restimulated in culture with

KLH those from mutant mice produced 3- to 20-fold more IFN-γ, confirming that the generation of Th1 cells was significantly increased in the absence of IL-10. Similar results have also been obtained with other antigens. The finding that IL-10 deficient mice exhibit exaggerated as well as prolonged DTH reactions supports the prediction that IL-10 production by Th2 T cells and macrophages plays an essential role in controlling cell-mediated immune responses.

In certain types of immune reactions suppression of the Th1 response by IL-10 is thought to favor the preferential generation of Th2 cells.[33] Th2 cells produce a specific set of cytokines (i.e. IL-4, IL-5, IL-6 and IL-10) which lead to strong humoral immunity and immediate hypersensitivity. Polarization of Th1- and Th2-type responses occurs frequently in parasitic infections and may result in resistance or susceptibility.[34] The outcome is dependent on the host's ability to generate the most effective type of immune response (Th1 versus Th2) necessary to eradicate a particular parasite. For example, resistance to infection with *Listeria monocytogenes* or *Nippostrongylus brasiliensis* (Nb) is dependent on the generation of a Th1 or a Th2 response, respectively.[35] To define more clearly the role of IL-10 in the generation of polarized immune responses to specific parasites, we compared the reactions of Nb-infected IL-10 deficient and control mice. As expected, normal mice exhibited a dominate Th2-type response (high IL-4, IL-5 and IgE production). The mutants not only developed a Th2 response, they also developed a strong Th1 response (high IFN-γ production).[7] These data demonstrate that, although IL-10 is not essential for the generation of functional Th2 cells, it is required to inhibit inappropriate (immunopathological?) Th1 responses to certain parasites.

CONCLUDING REMARKS

It is believed that inflammatory bowel disease (IBD) in man is due to an aberrant immune response to ordinary enteric antigens. In the past two decades, a tremendous effort has been made to understand the interactions between immune-reactive cells that lead to controlled immunity rather than immune-mediated pathogenesis as in the case of IBD. Despite the fact that numerous immunological alterations have been identified in the mucosal tissues of IBD patients,[36-40] it is doubtful that human studies will provide definitive answers to questions concerning genetic susceptibility factors, triggering agents and immunological defects that permit the initiation and amplification of the disease process.

Until recently, there was an absence of good experimental models for the two major forms of human IBD, Crohn's disease and ulcerative colitis. IL-10 deficient mice are only one of six mutants that have been generated by targeted gene inactivation and shown to develop chronic IBD. Because these mutants have had different components of their immune system "knocked out", it is not surprising that they manifest slightly different forms of disease. Mutants with disrupted genes for IL-2,[41] T-cell receptor (TCR) α, TCR β, TCR β × δ or class II MHC antigen[42] exhibit mucosal lesions in the colon that resemble ulcerative colitis. The disease pattern in IL-10 deficient mice housed in conventional facilities is distinctly different and consists of focal mucosal lesions in the small and large intestines. Upon aging, many IL-10 deficient mice (10-12 months old) show a transmural type of inflammation reminiscent of that observed in Crohn's disease. At this time, it is not known whether IL-10 deficient mice raised under SPF conditions will eventually develop a similar disease, or continue to exhibit focal lesions limited to the large intestine.[7]

Although it is not presumed that the cytokine and TCR knock-out mice are perfect models of IBD in man, it is anticipated that studies with these mutants will extend our understanding of the multiple regulatory events involved in the development of normal gut immunity. In fact, preliminary studies have already proven that uncontrolled inflammatory responses in these various mutant strains can be triggered by nonpathogenic microbial flora.[7,41,42] Moreover, the importance of genetic susceptibility factors has been demonstrated by the increased severity of disease in TCR α mutants inbred on 129/Sv and 129/Sv x C3H/He backgrounds relative to that observed in mutants inbred on the 129/Sv x Balb/c background.[42] Future studies are likely to settle other issues such as how T- and B-cell abnormalities lead to chronic, non-infectious intestinal inflammation and how imbalanced cytokine production contributes to the amplification and maintenance of the disease process.

Aside from the fact that IL-10 deficient mice may be useful in elucidating immunological defects that lead to IBD, studies with IL-10 deficient mice may also provide valuable insights into the basis of dysregulated immune-inflammatory reactions that contribute to disease susceptibility and immunopathology in other anatomical sites. For example, the inability of IL-10 deficient mice to establish an appropriately balanced immune response to agents that elicit DTH, contact hypersensitivity and irritant reactions invariable leads to prolonged and excessive inflammation resulting in unnecessary tissue damage and fibrosis. Therefore, information gained through studies with these mice may help understand how heightened inflammatory and Th1-like responses lead to chronic inflammatory and autoimmune diseases such as type 1 diabetes, arthritis, ankylosing spondylitis, scleroderma, uveitis, etc. It is of particular interest that IL-10 deficient mice exhibit abnormal immune responses in the skin. Skin diseases in humans with IBD are common and take many forms (eczema, erythema nodosum, pyoderma gangrenosum, etc.). Presently, we anticipate that studies with IL-10 deficient mice will help determine the potential efficacy of IL-10 or IL-10 antagonists in disease management.

ACKNOWLEDGMENTS

The DNAX Research Institute is supported by the Schering-Plough Corporation. Part of this work was supported by grants awarded to Dr. Klaus Rajewsky and Dr. Werner Müller by the Bundesministerium für Forschung und Technologie through the Genzentrum Köln by Land Nordrhein-Westfalen, and by the Fazit Foundation.

REFERENCES

1. Kühn R, Rajewsky K, Müller W. Generation and analysis of interleukin-4 deficient mice. Science 1991; 254:707-10.

2. Moore KW, Vieira P, Fiorentino DR et al. Homology of the cytokine synthesis inhibitory factor (IL-10) to the Epstein Barr virus gene BCRF1. Science 1990; 248:1230-34.

3. Go NF, Castle BE, Barrett R et al. Interleukin-10 (IL-10), a novel B cell stimulatory factor: unresponsiveness of X chromosome-linked immunodeficiency B cells. J Ex. Med 1990; 172:1625-31.

4. Fiorentino DF, Zlotnik A, Mosmann TR et al. IL-10 inhibits cytokine production by activated macrophages. J Immunol 1991; 147:3815-22.

5. de Waal-Malefyt R, Haanen J, Spits H et al. IL-10 inhibits cytokine synthesis by human monocytes: an autoregulatory role of IL-10 produced by monocytes. J Exp Med 1991; 174:1209-20.

6. Kim JM, Brannan CI, Copeland MG et al. Structure of the mouse IL-10 gene and chromosomal localization of the mouse and human genes. J Immunol 1992; 148:3618-23.

7. Kühn R, Löhler J, Rennick D et al. Interleukin-10-deficient mice develop chronic enterocolitis. Cell 1993; 75:263-74.

8. Howard M, Farrar J, Hilfiker M et al. Identification of a T cell-derived B cell growth factor distinct from interleukin 2. J Exp Med 1982; 155:914-23.

9. Lee F, Yokota T, Otsuka T et al. Isolation and characteri-zation of a mouse interleukin cDNA clone that expresses B-cell stimulatory factor 1 activities and T-cell- and mast-cell-stimulating activities. Proc Natl Acad Sci USA. 1986; 83:2061-65.

10. Paul WE, Ohara J. B cell stimulatory factor-1/interleukin-4. Annu Rev Immunol

1987; 5:429-59.

11. Ishida H, Hastings R, Kearney J et al. Continuous anti-interleukin-10 antibody administration depletes mice of Ly-1 B cells but not conventional B cells. J Exp Med 1992; 175:1213-20.

12. Bogdan C, Vodovotz Y, Nathan C. Macrophage deactivation by Interleukin-10 J Exp Med 1992; 174:1549-55.

13. de Waal-Malefyt R, Haanen J, Spits H et al. IL-10 and viral IL-10 strongly reduce antigen-specific human T cell proliferation by diminishing the antigen-presenting capacity of monocytes via downregulation of class II MHC expression. J Exp Med 1991; 174:915-24.

14. Ralph P, Nakoinz I, Sampson-Johannes A et al. IL-10, T lymphocyte inhibitor of human blood cell production of IL-1 and tumor necrosis factor. J Immunol 1992; 148:808-14.

15. Hsu D-H, Moore KW, Spits H. Differential effects of interleukins-4 and -10 on interleukin-2-induced interferon-γ synthesis and lymphokine-activated killer activity. Int Immunol 1992; 4:563-69.

16. Fiorentino DF, Bond MW, Mosmann TR. Two types of mouse helper T cell. IV. Th2 clones secrete a factor that inhibits cytokine production by Th1 clones. J Exp Med 1989; 170:2081-95.

17. Fiorentino DF, Zlotnik A, Vieira P et al. IL-10 acts on the antigen-presenting cell to inhibit cytokine production by Th1 cells. J Immunol 1991; 146:3444-51.

18. Taga K, Tosato G. IL-10 inhibits T cell proliferation and IL-2 production. J Immunol 1992; 148:1143-48.

19. Ding L, Shevach EM. IL-10 inhibits mitogen-induced T cell proliferation by selectively inhibiting macrophage costimulatory function. J Immunol 1992: 148:3133-39.

20. O'Garra A, Stapleton G, Dhar V et al. Production of cytokines by mouse B cells: B lymphomas and normal B cells produce Interleukin-10. Int Immunol 1990; 2: 821-32.

21. O'Garra A, Chang R, Go N et al. Ly-1 B (B-1) cells are the main source of B-cell-derived IL-10. Eur J Immunol 1992; 22:711-17.

22. Yssel H, de Waal Malefyt R, Roncarolo M-G et al. IL-10 is produced by subsets of human CD4⁺ T cell clones and peripheral blood T cells. J Immunol 1992; 149: 2378-84.

23. Lin TZ, Svetic A, Ganea D et al. Cytokines in NZB CD5⁺ B clones. Ann NY Acad Sci 1992; 651:581-83.

24. Hisatsun T, Minai Y, Nishisima KI et al. A suppressive lymphokine derived from Ts clone 13G2 is IL-10. Lymphokine Cytokine Res 1992; 11:87-93.

25. Benjamin D, Knobloch TJ, Dayton MA. Human B-cell interleukin-10: B-cell lines derived from patients with acquired immunodeficiency syndrome and Burkitt's lymphoma constitutively secrete large quantities of interleukin-10. Blood 1992; 80:1289-98.

26. Rivas JM, Ullrich SE. Systemic suppression of delayed-type hypersensitivity by supernatants from UV-irradiated keratinocytes. An essential role for keratinocyte-derived IL-10. J Immunol 1992; 149:3865-71.

27. Enk AJ, Katz SI. Identification and induction of keratinocyte-derived IL-10. J Immunol 1992; 149:92-95.

28. Gocinski BL, Tigelaar RE. Roles of CD4⁺ and CD8⁺ T cells in murine contact sensitivity revealed by in vivo monoclonal antibody depletion. J Immunol 1990; 144: 4121-28.

29. Piguet PF, Grau GE, Vassalli P. Subcutaneous perfusion of tumor necrosis factor induces local proliferation of fibroblasts, capillaries, and epidermal cells, or massive tissue necrosis. Am J Pathol 1990; 136: 103-09.

30. Piguet PF, Grau, GE, Hauser C et al. Tumor necrosis factor is a critical mediator in hapten-induced irritant and contact hypersensitivity reactions. J Exp Med 1991; 173:673-79.

31. Fong TAT, Mosmann TR. The role of IFN-gamma in delayed-type hypersensitivity mediated by Th1 clones. J Immunol 1989; 143:2887-93.

32. Issekutz TB, Stoltz JM, Van der Meide P. Lymphocyte recruitment in delayed-type hypersensitivity. The role of IFN-g. J Immunol 1988; 140:2989-93.

33. Mosmann TR, Moore, KW. The role of IL-10 in crossregulation of Th1 and Th2 responses. Immunol Today 1991; 12: A49-53.

34. Sher A, Coffman RL. Regulation of immunity to parasites by T cells and T cell-derived cytokines. Ann Rev Immunol 1992; 10:385-409.

35. Urban JF, Madden KB, Svetic A et al. The importance of Th2 cytokines in protective immunity to nematodes. Immunol Rev 1992; 127:205-20.

36. Shanahan F. Pathogenesis of ulcerative colitis. The Lancet 1993; 342:407-11.

37. Schreiber S, MacDermott RP, Raedler A et al. Increased activation of isolated intestinal lamina propria mononuclear cells in inflammatory bowel disease. Gastroenterology 1991; 101:1020-30.

38. Mayer L, Eisenhardt D. Lack of induction of suppresser T cells by intestinal epithelial cells from patients with inflammatory bowel disease. J Clin Invest 1990; 86:1255-60.

39. Mullin GE, Lazenby AJ, Harris ML et al. Increased interleukin-2 messenger RNA in the intestinal mucosal lesions of Crohn's disease but not ulcerative colitis. Gastroenterology 1992; 102:1620-27.

40. Kaulfersch W, Fiocchi C, Waldman TA. Polyclonal nature of the intestinal mucosal lymphocyte populations in inflammatory bowel disease. A molecular genetic evaluation of the immunoglobulin and T-cell antigen receptors. Gastroenterology 1988; 95:364-70.

41. Sadlack B, Merz H, Schorle H et al. Ulcerative colitis-like disease in mice with a disrupted interleukin-2 gene. Cell 1993; 75:253-61.

42. Mombaerts P, Mizoguchi E, Grusby MJ et al. Spontaneous development of inflammatory bowel disease in T cell receptor mutant mice. Cell 1993; 75:275-82.

A PHASE I STUDY OF INTERLEUKIN-10 IN HEALTHY HUMANS: SAFETY AND EFFECTS ON CYTOKINE PRODUCTION

Amy E. Chernoff, Eric V. Granowitz, Leland Shapiro,
Edouard Vannier, Gerhard Lonnemann, Jonathan B. Angel,
Scott F. Orencole, Jeffrey S. Kennedy, Xi-Xian Zhang,
Hei-De Wen, Ellen C. Donaldson, Elaine Radwanski,
David L. Cutler, Sheldon M. Wolff, and Charles A. Dinarello

INTRODUCTION

Interleukin-10 (IL-10) has multiple in vitro actions on different cell types including thymocytes,[1] T and B lymphocytes,[2-4] monocytes,[5] neutrophils,[6] and mast cells.[7] When cultured with T lymphocytes, IL-10 suppresses IL-2 and IFN-γ production and inhibits mitogen-induced T-cell proliferation.[8,9] On the other hand, IL-10 stimulates B-cell growth and immunoglobulin (Ig) production.[4] In monocytes, IL-10 inhibits synthesis and gene expression for IL-1, TNF, IL-6, IL-8, and colony stimulating factors.[5,8] Similar effects have been observed in neutrophils.[6] IL-10 also prevents macrophage cytotoxic activity by inhibiting both cytokine and nitric oxide production.[10]

Because of its ability to inhibit production of IL-1, TNF and IL-8, cytokines involved in acute and chronic inflammatory processes, IL-10 may be useful in a number of conditions such as sepsis, chronic arthritis and inflammatory bowel disease. In animal models of sepsis, IL-10, given prior to, concurrently with, or soon after gram-negative bacterial endotoxin, or staphylococcal enterotoxin B, prevents TNFα production, hypothermia, and death.[11-13] There may be a role for endogenous IL-10 production

Interleukin-10, edited by Jan E. deVries and René de Waal Malefyt.

in disease. Interleukin-10 deficient mice develop chronic enterocolitis[14] and humans with inflammatory bowel disease have been shown to have low levels of circulating IL-10.[15] In addition, IL-10 suppresses cell-mediated immunity via its effects on T-helper type 1 cells and may therefore be useful in the treatment of transplant rejection. In patients with severe combined immunodeficiency who underwent bone marrow transplantation, tolerance to HLA mismatch was associated with high circulating levels of IL-10.[16]

We performed a Phase I study of intravenous IL-10 in healthy male volunteers. We examined the safety and serum concentrations after a single bolus injection of escalating doses of IL-10. In addition, cytokine production in peripheral blood mononuclear cells (PBMC) and whole blood was measured following IL-10 administration. We also examined the effect of IL-10 on neutrophil superoxide generation.

MATERIALS AND METHODS

VOLUNTEER SELECTION

Eighteen healthy, nonsmoking male volunteers between 18 and 40 years of age were enrolled after giving informed consent. The study protocol and consent forms were approved by the Human Investigations Review Committee of the New England Medical Center and Tufts University School of Medicine. Eligibility requirements included the absence of underlying disease as assessed by history, physical examination, and laboratory studies. These included complete blood count (hemoglobin, total and differential white blood cell count, and platelets), blood chemistries (sodium, potassium, chloride, bicarbonate, glucose, blood urea nitrogen, creatinine, uric acid, calcium, phosphate, total protein, albumin, alkaline phosphatase, total bilirubin, alanine aminotransferase, aspartate aminotransferase, gamma glutamyl-transpeptidase, lactate dehydrogenase, cholesterol, and triglycerides), urinalysis (macroscopic and microscopic), viral serologies (human immunodeficiency virus type 1, hepatitis B, and hepatitis C) electrocardiogram, and chest radiograph. Volunteers abstained from using oral cyclooxygenase inhibitors or anti-histamines during the 2 weeks prior to the study. Volunteers were excluded if they had used pentoxifylline within 3 months of the study.

STUDY DESIGN

This was a randomized, double-blind, placebo-controlled, parallel-group study. Volunteers were admitted to the Clinical Study Unit at the New England Medical Center for 60 hours. After a 10-hour overnight fast, volunteers received either human recombinant IL-10 (Schering-Plough, Kenilworth, NJ) or placebo (a lyophilized formulation without drug) as an intravenous bolus injection given over 30 seconds into the antecubital fossa. Venous blood samples during the initial 6 hours following drug administration were obtained from an indwelling intravenous catheter in the contralateral arm. The doses of IL-10 were 1 µg/kg (4 subjects), 10 µg/kg (4 subjects), and 25 µg/kg of body weight (4 subjects). Six volunteers received placebo. One subject in the placebo group was eliminated from all analyses due to numerous unsuccessful attempts at venous access.

CLINICAL AND LABORATORY EVALUATION

Vital signs (temperature, pulse, blood pressure and respiratory rate) were recorded just prior to and 0.25, 0.5, 1, 2, 4, 8, 12, 24, 36 and 48 hours after drug administration. Volunteers were questioned during the study period of potential drug-related side effects. A complete blood count was obtained immediately before and 3, 6, 24, 48 and 96 hours after the injection. Blood chemistries and urinalyses were obtained upon admission and 3, 24 and 48 hours after drug administration. Blood samples for Coombs test, reticulocyte count and serum haptoglobin were collected immediately before and 24, 48 and 96 hours after study drug injection. Plasma fibrinogen, C3, C4 and total hemolytic complement were obtained immediately before and 3 and 12 hours after IL-10. Serum IgG, IgA and IgM were quantitated just prior to and 96 hours after dosing. An electrocardiogram was performed upon admission and 2 and 48 hours post-treatment.

SERUM IL-10 CONCENTRATIONS

Blood was collected in sterile, vacuum blood collection tubes (Beckton-Dickinson, Rutherford, NJ) immediately prior to and 2, 5, 10, 20, 30, 45 minutes and 1, 1.5, 2, 2.5, 3, 4, 5, 6, 8, 12, 16 and 24 hours after drug administration. Blood was allowed to clot at room temperature for 20 minutes and then centrifuged for 15 minutes at 4°C. Serum was separated and stored at -70°C. Serum samples were later assayed for IL-10 concentration using an enzyme-linked immunosorbent assay with a limit of detection of 100 pg/ml.

ANTIBODIES TO IL-10

Blood was collected in sterile, vacuum blood collection tubes 12 hours prior to and 21 days after injection of IL-10. The blood was allowed to clot at room temperature for 30 minutes and then centrifuged at 4°C for 15 minutes. Serum was separated and stored at -70°C. Subsequently, samples were analyzed for antibodies to IL-10 by ELISA. Serial dilutions of serum samples were added to wells of a microtiter plate pre-coated with human IL-10. After overnight incubation, the presence of IL-10 antibodies was detected by addition of biotin-labeled protein A followed by streptavidin conjugated to horseradish peroxidase. The results of each sample in this ELISA were compared with the results from pooled normal human serum. Samples were considered positive if 1) the ratio of optical density from the serum sample to optical density from the normal human serum was greater than or equal to 1.0 and 2) the ratio of optical density from the serum sample after IL-10 dosing to optical density from the serum sample prior to dosing from the same volunteer was greater than or equal to 2.0.

ISOLATION OF PERIPHERAL BLOOD MONONUCLEAR CELLS (PBMC)

Blood was drawn into heparinized syringes (10 U/ml) immediately before and 3, 6, 24 and 48 hours after injection of IL-10 or saline. PBMC were isolated by centrifugation through Ficoll (Sigma Chemical Co., St. Louis, MO) and Hypaque (90%, Winthrop Laboratories, New York, NY). PBMC were washed twice in 0.9% sodium chloride and resuspended.

ISOLATION OF NEUTROPHILS

After density gradient centrifugation through Ficoll and Hypaque, the PBMC and plasma layers were removed leaving the neutrophil-red cell pellet. Neutrophils were mixed with saline (1:1) and sedimented in 3% dextran. Neutrophils were then washed in saline and remaining red blood cells lysed using water (4°C) followed by hypertonic NaCl (9.0%) to return to isosmotic conditions. The neutrophils were washed and resuspended in cold Hanks' balanced salt solution (HBSS) at a concentration of 107 cells/ml.

WHOLE BLOOD CYTOKINE PRODUCTION

One milliliter of either RPMI or LPS (from *E. coli* 055:B5, Sigma) diluted in RPMI was aliquoted into polypropylene tubes and 2.5 ml of heparinized blood were added (final concentration LPS, 10 ng/ml). Blood was then incubated for 24 hours at 37°C in 5% CO_2. After incubation, the supernatant plasma was transferred into sterile 1.5 ml microfuge tubes and centrifuged for 2 minutes at 10,000g. The resulting platelet-poor plasma was removed and frozen at -70°C in polypropylene tubes.

INDUCTION OF GM-CSF PRODUCTION IN PBMC

Phorbol 12-myristate 13-acetate (PMA) was purhased from Sigma and diluted in DMSO at 1 mg/ml. Phytohemagglutinin (PHA) was purchased form Difco Laboratories, Detroit, Michigan. PBMC were resuspended at a concentration of 5×10^6 cells/ml in ultrafiltered RPMI culture medium 1640 (Sigma) supplemented with 10 mM L-glutamine, 24 mM NaHCO₃ (Mallinckrodt, Paris, KY), 10 mM HEPES (Sigma), 100 U/ml penicillin and 100 µg/ml streptomycin (Gibco Laboratories, Grand Island, NY). 2% v/v heat-inactivated human AB serum was added. Cells (750 µl) were aliquoted into 12 x 75 mm round-bottom polypropylene tubes (Falcon, Becton-Dickinson, Lincoln Park, NJ) and 750 µl of either RPMI or PHA and PMA in RPMI (final concentration PHA 10 µg/ml, PMA 100 ng/ml) was added. PBMC were then incubated for 24 hours at 37°C in a humidified atmosphere containing 5% CO_2. The cultures were frozen at -70°C. Prior to assay, cells were subjected to three freeze-thaw cycles for total cytokine recovery.

SUPEROXIDE RELEASE FROM NEUTROPHILS

Superoxide was measured by the reduction of ferricytochrome C to ferrocytochrome C. Neutrophils (final concentration 5 x 10⁶/ml) were aliquoted into 1.5 ml Eppendorf tubes with 1.2 mg/ml cytochrome C (Sigma) and incubated for 15 minutes in a 37°C water bath under the following conditions: (1) with 100 µg/ml superoxide dismutase (SOD, Sigma) in HBSS, (2) with 500 ng/ml PMA in HBSS, (3) with SOD (100 µg/ml) plus PMA (500 ng/ml) in HBSS and (4) with HBSS alone. Following incubation, tubes were centrifuged for 1 minute at 10,000 g to pellet the cells. The supernatant was transferred to a 96-well microplate. Cytochrome C reduction was determined by optical density at 550 nm.

CYTOKINE ASSAYS

Plasma from whole blood incubation was analyzed by specific radioimmunoassays for IL-1β,[17] TNFα,[18] IL-6,[19] IL-8[20] and IL-1Ra.[21] Sensitivities of the radioimmunoassays for each cytokine were as follows: IL-1β, 40 pg/ml; TNFα, 80 pg/ml; IL-6, 40 pg/ml; IL-8, 80 pg/ml; and IL-1Ra, 80 pg/ml. GM-CSF levels in PBMC were determined by an ELISA (Genzyme Inc.,

Cambridge, MA). The limit of detection for GM-CSF was 10 pg/ml.

STATISTICAL ANALYSIS

IL-10 serum levels for the three dosing groups are expressed as the mean ± SEM. Cytokine data obtained at 3, 6, 24 and 48 hours for each volunteer were expressed as a percentage of the baseline value for that subject at time zero. Percentages are shown as the mean ± SEM. Statistical analyses were based on a comparison of data at times 3, 6, 24 and 48 hours with baseline values at time zero. Statistics were performed by analysis of variance (ANOVA) using Fisher's least significance difference.

RESULTS

CLINICAL AND LABORATORY EVALUATION

Table 17.1 shows the times of clinical and laboratory evaluation. Clinical studies revealed that there were no adverse symptoms reported by or observed in volunteers injected with IL-10. In addition, there were no clinically relevant differences in blood chemistries, urinalyses, hemoglobin concentrations, platelet counts, coagulation parameters, complement components, immunoglobulin concentrations, or electrocardiograms between any dose-level group receiving IL-10 and the placebo-injected volunteers.

One volunteer developed a faint, erythematous, maculopapular rash on his back and arms 3 days after administration of 10 µg/kg IL-10. Five days after injection, the rash resolved. The volunteer reported a previous history of detergent allergy. The rash was present in only those skin areas that had contact with the hospital sheets. A second volunteer had mild first degree AV block (PR interval 220 msec) prior to injection. Two hours after administration of 25 µg/kg IL-10, his PR internal increased to 233 msec. However, 48 hours after injection his PR interval had returned to normal (190 msec).

Table 17.1. Schedule of studies in volunteers receiving IL-10

Evaluation	Time (after bolus injection of IL-10)																							
	-12 hrs	0	2 min	5 min	10 min	20 min	30 min	45 min	1 hr	1.5 hrs	2 hrs	2.5 hrs	3 hrs	4 hrs	5 hrs	6 hrs	8 hrs	12 hrs	16 hrs	24 hrs	36 hrs	48 hrs	96 hrs	21 days
Clinical evaluation																								
Symptoms and vital signs	x	x			x		x		x	x	x			x			x	x		x	x	x		
Injection site evaluation		x			x		x		x		x			x			x	x		x	x	x		
Electrocardiogram	x										x											x		
Clinical laboratory																								
CBC, diff, platelets	x	x														x				x		x	x	
Blood chemistries	x												x							x		x		
Urinalysis	x												x							x		x		
C3, C4, CH50		x											x										x	
Serum IgG,IgM,IgA		x																						
Coombs test		x																						
Reticulocyte count		x																		x				
Haptoglobin		x																		x				
Fibrinogen		x											x					x						
Serum IL-10 levels		x	x	x	x	x	x	x	x	x	x	x	x	x	x	x	x	x	x	x				
IL-10 Antibody	x																							x
Immunologic tests																								
Stimulus-induced cytokines: PBMC, whole blood		x			x								x			x				x		x		
Neutrophil superoxide generation		x			x								x			x				x		x		

SERUM IL-10 CONCENTRATIONS

Serum IL-10 levels in the placebo-injected controls were less than 0.1 ng/ml at each time point. One subject in the IL-10 1 µg/kg group was eliminated from analysis due to inhibitors present in his blood which interfered with the IL-10 ELISA. One volunteer in the 25 µg/kg group had 0.2 ng/ml of IL-10 in his serum prior to the injection. This subject's level returned to 0.2 ng/ml 16 and 24 hours after injection but never became undetectable. The other three volunteers in that group had undetectable levels of IL-10 24 hours after the injection. The mean peak serum concentrations of IL-10 after bolus injection were 14 ± 1.3 ng/ml for the 1 µg/kg dose group, 208 ± 20.1 ng/ml for the 10 µg/kg dose group, and 505 ± 22.3 ng/ml for the 25 µg/kg dose group. Peak concentrations were reached 2-5 minutes following injection. Mean IL-10 serum concentrat-ions for each group 3, 6 and 24 hours after injection are shown in Table 17.2.

ANTIBODIES TO IL-10

Antibodies to IL-10 prior to the start of the study and twenty one days after injection were not detected.

EFFECT OF IL-10 ON WHOLE BLOOD CYTOKINE PRODUCTION

There were no significant differences in whole blood production of IL-1β, TNF-α, IL-6, IL-1Ra, or IL-8 prior to the injection of any dose of IL-10 and the placebo-injected controls. In addition, in the absence of LPS stimulation, spontaneous whole blood production of these cytokines 3, 6, 24, or 48 hours after the injection was not observed.

In contrast, marked inhibition of LPS-induced IL-6 production occurred in all IL-10 dose groups (Fig 17.1). In the low dose group (1 µg/kg) an inhibition of 60% of baseline was observed 3 hours after the injection but production returned to pre-injection levels by 6 hours. In the 10 µg/kg group, there was a 90% reduction in LPS-induced IL-6 production at 3 hours (P<0.001). Inhibition persisted in this group for 6 hours (69%, P<0.001). In the high dose group (25 µg/kg), inhibition was maximal (91-95%, P<0.001) at 3 and 6 hours and persisted for 48 hours after the injection (47%, P<0.001). Other cytokines (IL-1β, TNFα) were similarly suppressed (data not shown).

EFFECT OF IL-10 ON GM-CSF PRODUCTION IN PBMC

There were no significant changes in GM-CSF production by PBMC in response to stimulation by PHA (10 µg/ml) plus PMA (100 ng/ml) between any of the IL-10 dose groups and the placebo-injected controls.

SUPEROXIDE RELEASE FROM NEUTROPHILS

There were no significant differences in neutrophil superoxide generation between any IL-10 dose group and the placebo-injected controls.

Table 17.2. IL-10 levels in serum

IL-10 Dose	Hours after IL-10 Injection	IL-10 Serum Concentration (ng/ml) Mean ± SEM
1 µg/kg	3	0.16 ± 0.04
	6	< 0.1
	24	< 0.1
10 µg/kg	3	6.27 ± 1.4
	6	0.8 ± 0.25
	24	< 0.1
25 µg/kg	3	22.93 ± 3.15
	6	3.63 ± 0.83
	24	0.12 ± 0.03

DISCUSSION

IL-10 inhibits monocyte and neutrophil synthesis of IL-1, TNF, IL-6 and IL-8 and suppresses cell mediated immunity via its effects on T cells. Consequently, there are several potential therapeutic uses for IL-10 in acute and chronic inflammatory diseases as well as during transplant rejection. However, before IL-10 can be tested clinically, it is important to demonstrate the safety of IL-10 administration in humans.

In the present study, 18 healthy HIV-negative males received a single bolus injection of IL-10 (1, 10 and 25 µg/kg) or placebo. There were no adverse symptoms or signs reported or observed. In addition, there were no clinically significant differences in the hemoglobin concentrations, platelet counts or blood chemistries between the groups receiving IL-10 and the placebo-injected controls. One volunteer developed a rash 3 days after injection of 10 µg/kg IL-10. This subject had a known history

of detergent allergy and the rash was present only on areas exposed to the sheets. It is probable that this rash represents an allergic response independent of IL-10. A second volunteer developed a slight increase in his baseline 1st degree AV block (PR 220-233 msec) 2 hours after injection of 25 µg/kg IL-10. However, an electrocardiogram performed 48 hours following injection revealed that he no longer had 1st degree AV block (PR 190). Given the variable PR interval in this volunteer, it is unclear whether IL-10 has an effect on AV node conduction. In summary, administration of IL-10 as a single bolus injection appears safe.

When added to B-cell cultures in vitro, IL-10 stimulates growth and Ig production.[4] In fact, when IL-10 was added to B cells from patients with common variable immunodeficiency, a disease characterized by inadequate B-cell production of immunoglobulins, synthesis of IgG, IgA and IgM was stimulated.[22] In our volunteers,

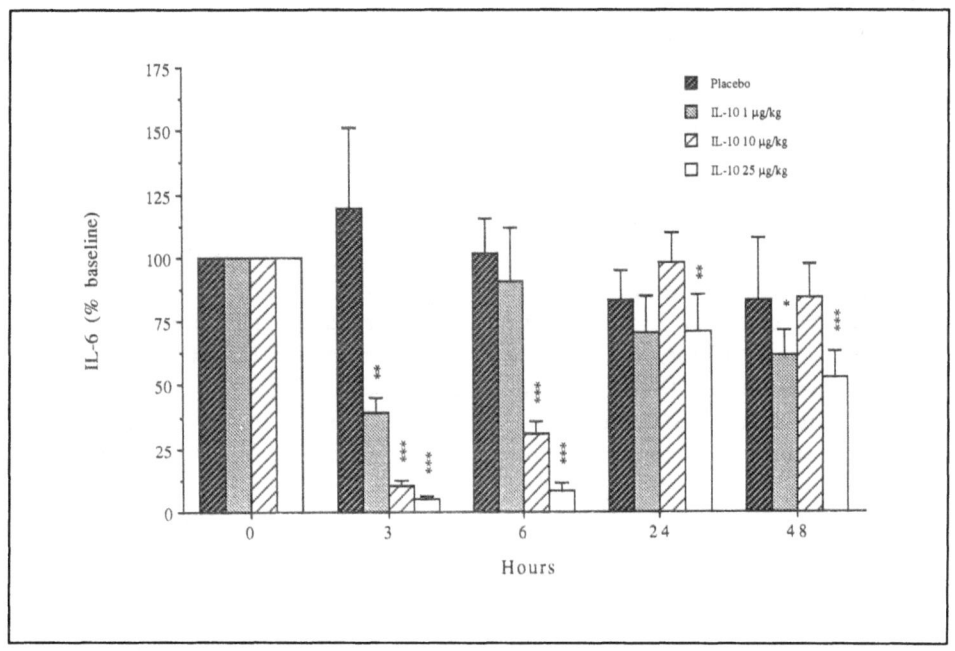

Fig. 17.1. Effect of in vivo IL-10 on IL-6 production in whole blood stimulated with LPS. Heparinized blood was collected from volunteers immediately before and 3, 6, 24, and 48 hours after a bolus injection of IL-10 or saline and then stimulated with LPS (10 ng/ml) for 24 hours. For each volunteer, LPS-induced IL-6 production was measured by specific radioimmunoassay. The data are expressed as a percentage of IL-6 produced at time zero. Data are depicted as the mean ± SEM. Differences were analyzed by ANOVA (* = p<0.05, ** = p<0.01, *** = p < 0.001)

however, quantitative serum immunoglobulin levels in each of the groups receiving IL-10 were unchanged 96 hours after IL-10 administration. The most likely reason for these disparate results is that the B cells in the in vitro experiments were preactivated with anti-CD40 or *Staphylococcus aureus* Cowan strain. In the present study, B-cell activators were not given to the volunteers. The possibility remains, however, that if IL-10 is given to patients with deficient Ig production and/or if given repeatedly or in higher doses, it may increase immunoglobulin production by B cells in vivo.

We found no spontaneous production of IL-1, TNF, IL-6, IL-8, or IL-1Ra in whole blood removed from the volunteers 3, 6, 24, or 48 hours after IL-10 injection and subsequently incubated for 24 hours. These data confirm in vitro studies that IL-10 does not induce the synthesis of cytokines by monocytes, T lymphocytes, or neutrophils. Although IL-10 has been shown to induce in vitro production of IL-6 by endothelial cells[23] we did not observe an increase in plasma levels of IL-6 following injection of IL-10.

We found no inhibition of GM-CSF production in stimulated PBMC from volunteers who had received IL-10 at any dose. This contrasts with prior observations that IL-10 inhibits GM-CSF production when added to PBMC cultures stimulated with PHA or anti-CD3.[8] One possible explanation for this disparity is that in the in vitro experiments, PBMC are exposed to high concentrations of IL-10 for prolonged periods of time. PBMC from volunteers in our study are exposed to low doses of IL-10 in vivo for a short period of time. In fact, even in the highest dose group in our study, mean serum IL-10 levels were only 3.62 ng/ml 6 hours after the injection and undetectable by 24 hours. It is possible that longer exposure to higher levels of IL-10 in vivo are necessary for IL-10 mediated inhibition of GM-CSF production in PBMC.

When administered in vivo, IL-10 affected subsequent cytokine production.

When stimulated ex vivo with LPS, whole blood synthesis of IL-6 was significantly inhibited in each group receiving IL-10. In the 1 µg/kg group, 60% inhibition was observed 3 hours after the injection. In the group receiving 10 µg/kg, there was a 90% reduction at 3 hours and a 69% reduction at 6 hours. In the volunteers injected with 25 µg/kg, 91-95% inhibition occurred at 3 and 6 hours and persisted for 48 hours (47%) after the injection. We considered the possibility that the inhibition of IL-6 production may be a consequence of the in vitro activity of residual IL-10 present in the whole blood samples. However, our data suggest that most of the inhibition cannot be explained by this mechanism. In the low dose group, 60% inhibition was observed at 3 hours when mean serum concentration of IL-10 was only 0.13 ng/ml. In the 10 µg/kg group, 69% inhibition of IL-6 production occurred at 6 hours when mean serum concentrations of IL-10 were only 0.8 ng/ml. In our laboratory, we have observed that greater than 5 ng/ml of in vitro IL-10 are necessary for any suppression of LPS-induced IL-6 production in whole blood (data not shown). Furthermore, suppression of IL-6 synthesis persisted at 48 hours (47%) in the high dose group despite undetectable levels of IL-10 in serum by 24 hours after the injection. These data also have implications regarding the future use of IL-10 for therapy. It is likely that daily treatment will not be necessary since the biological effects of IL-10 on cytokine synthesis are prolonged for at least 48 hours.

IL-6 is an endogenous pyrogen,[24] a potent stimulator of B-cell growth and immunoglobulin production,[25] and plays an important role in the acute phase response.[26] IL-6 has thus been implicated in the pathogenesis of a number of clinical conditions including malignancies and autoimmune diseases. It is produced by and acts as a growth factor for myeloma and plasmacytoma cells and anti-IL-6 has been shown to inhibit the growth of these cells.[27,28] A role for IL-6 in autoimmune disease has been proposed due to its ability to induce polyclonal B-cell activation. In support of this, high levels of IL-6 have been found in serum and synovial fluid from the joints of patients with rheumatoid arthritis.[29] This may explain the infiltration of plasma cells into the joints, the elevation of acute phase proteins, and the systemic autoimmune phenomena seen in these patients. Patients with atrial myxomas often have associated autoimmune phenomena and examination of these tumors has revealed increased production of IL-6.[30] IL-6 has also been implicated in the pathogenesis of Castelman's disease, a disease characterized by fever, benign lymphadenopathy, hepato-splenomegaly, hypergammaglobulinemia and elevation in acute phase proteins. Lymph nodes excised from these patients have been shown to produce large amounts of IL-6 and serum IL-6 levels are elevated.[31] Treatment of Castelman's patients with anti-IL-6 antibodies has recently been reported to be effective.[32] IL-10, by inhibiting IL-6 production, may therefore be useful in treating multiple myeloma, plasmacytomas, autoimmune diseases and Castelman's disease. One must be cautious however when considering treating B-cell mediated diseases with IL-10 since like IL-6, IL-10 stimulates B cells in vitro. Clearly this warrants further investigation.

In this study we have shown that a single injection of recombinant human IL-10 in healthy humans is safe. In addition, we observed profound effects on ex vivo cytokine production. It is reasonable to now study this cytokine in human disease states.

ACKNOWLEDGMENTS

This work was supported by National Institutes of Health grants AI-07329, AI-15614, MO1RR00054-033, and RO1-AI-33290 (EVG) and by Schering-Plough Research Institute. Gerhard Lonnemann is supported by DFG Lo 535/1-1.2. The authors wish to thank Timothy Cummings, Donna Keany, Patricia M. Noga, Aubrey E. Boyd and the staff of the Clinical Study Unit of the New England Medical Center. We are grateful to Dr. Sheila Jacobs and Ron Sabo of Schering Plough Research Institute for their assistance.

REFERENCES

1. MacNeil IA, Suda T, Moore KW et al. IL-10, a novel growth cofactor for mature and immature T cells. J Immunol 1989; 145:4167-73.
2. Del Prete G, De Carli M, Almerigogna F et al. Human IL-10 is produced by both type 1 helper (Th1) and type 2 helper (Th2) T cell clones and inhibits their antigen-specific proliferation and cytokine production. J Immunol 1993; 150:353-60.
3. Fiorentino DF, Zlotnik A, Vieira P et al. IL-10 acts on the antigen-presenting cell to inhibit cytokine production by TH1 cells. J Immunol 1991; 146:3444-51.
4. Rousset F, Garcia E, Defrance T et al. Interleukin 10 is a potent growth and differentiation factor for activated human B lymphocytes. Proc Natl Acad Sci USA 1992; 89:1890-93.
5. de Waal Malefyt R, Abrams J, Bennett B et al. Interleukin 10 (IL-10) inhibits cytokine synthesis by human monocytes: An autoregulatory role of IL-10 produced by monocytes. J Exp Med 1991; 174:1209-20.
6. Cassatella MA, Meda L, Bonora S et al. Interleukin 10 (IL-10) inhibits the release of proinflammatory cytokines from human polymorphonuclear leukocytes. Evidence for an autocrine role of tumor necrosis factor and IL-1β in mediating the production of IL-18 triggered by lipopolysaccharide. J Exp Med 1993; 178:2207-11.
7. Thompson-Snipes L, Dhar V, Bond W et al. Interleukin 10: A novel stimulatory factor for mast cells and their progenitors. J Exp Med 1991; 173:507-10.
8. Vieira P, De Waal Malefyt R, Dang MN et al. Isolation and expression of human cytokine synthesis inhibitory factor cDNA clones: Homology to Epstein-Barr virus open reading frame BCRFI. Proc Natl Acad Sci USA 1991; 88:1172-76.
9. Taga K, Tosato G. IL-10 inhibits T cell proliferation and IL-2 production. J Immunol 1992; 148:1143-48.
10. Bogdan C, Vodovotz Y Nathan C. Macrophage deactivation by IL-10. J Exp Med

1991; 174:1549-55.

11. Bean AGD, Freiberg RA, Andrade S et al. Interleukin 10 protects mice against staphylococcal enterotoxin B-induced lethal shock. Infect Immun 1993; 61:4937-39.

12. Gerard C, Bruyns C, Marchant A et al. Interleukin 10 reduces the release of tumor necrosis factor and prevents lethality in experimental endotoxemia. J Exp Med 1993; 177:547-50.

13. Howard M, Muchamuel T, Andrade S et al. Interleukin 10 protects mice from lethal endotoxemia. J Exp Med 1993; 177:1205-8.

14. Kuhn R, Lohler J, Rennick D et al. Interleukin-10-deficient mice develop chronic enterocolitis. Cell 1993; 75:263-74.

15. Ishida H. Clinical implications of IL-10 in patients with immune and inflammatory diseases. Lymphokine Cytokine Res (abstr) 1993; 12:344.

16. Bacchetta R, Bigler M, Touraine J et al. High levels of interleukin 10 production in vivo are associated with tolerance in SCID patients transplanted with HLA mismatched hematopoietic stem cells. J Exp Med 1994; 179:493-502.

17. Lisi PJ, Chu CW, Koch, GA et al. Development and use of a radioimmunoassay for human interleukin 1ß. Lymphokine Res 1987; 7:75-84.

18. van der Meer JWM, Endres S, Lonnemann G et al. Concentrations of immunoreactive human tumor necrosis factor-α produced by human mononuclear cells in vitro. J Leukoc Biol 1988; 43:216-23.

19. Schindler R, Mancilla J, Endres S et al. Correlations and interactions in the production of interleukin-6 (IL-6), IL-1, and tumor necrosis factor (TNF) in human blood mononuclear cells: IL-6 suppresses IL-1 and TNF. Blood 1990; 75:40-47.

20. Porat R, Poutsiaka DD, Miller LC et al. Interleukin-1 (IL-1) receptor blockade reduces endotoxin and Bor*relia burgdorferi*-stimulated IL-8 synthesis in human mononuclear cells. FASEB J 1992; 6:2482-2486.

21. Poutsiaka DD, Clark BD, Vannier E et al. Production of interleukin-1 receptor antagonist and interleukin-1β by peripheral blood mononuclear cells is differentially regulated. Blood 1991; 78:1275-81.

22. Zielen S, Bauscher P, Hofmann D et al. Interleukin 10 and immune restoration in common variable immunodeficiency [letter]. Lancet 1993; 342:750-1.

23. Sironi M, Munoz C, Pollicino T et al. Divergent effects of Interleukin-10 on cytokine production by mononuclear phagocytes and endothelial cells. Eur J Immunol 1993; 23:2692-95.

24. Helle M, Brakenhoff JP, De GER et al. Interleukin 6 is involved in Interleukin 1-induced activities. Eur J Immunol 1988; 18:957-59.

25. Hirano T, Taga T, Nakano N et al. Purification to homogeneity and characterization of human B cell differentiation factor (BCDF or BSFp-2). Proc Natl Acad Sci USA 1985; 82:5490-94.

26. Gauldie J, Richards C, Harnish D et al. Interferon β2/B-cell stimulatory factor type 2 shares identity with monocyte-derived hepatocyte-stimulating factor and regulates the major acute phase protein response in liver cells. Proc Natl Acad Sci USA 1987; 84:7251-55.

27. Kawano M, Hirano T Matsuda T, Taga T et al. Autocrine generation and requirement of BSF-2/IL-6 for human multiple myelomas. Nature 1988; 332:83-5.

28. Nordan RP, Potter M. A macrophage-derived factor required by plasmacytomas for survival and proliferation in vitro. Science 1986; 233:566-69.

29. Houssiau FA, Devogelaer J-P, Van Damme J et al. Interleukin-6 in synovial fluid and serum of patients with rheumatoid arthritis and other inflammatory arthritides. Arthritis Rheum 1988; 31:784-88.

30. Hirano T, Taga T, Kiyoshi Y et al. Human B-cell differentiation factor defined by an anti-peptide antibody and its possible role in autoantibody production. Proc Natl Acad Sci USA 1987; 84:228-31.

31. Yoshizaki K, Matsuda T, Nishimoto N et al. Pathological significance of interleukin 6 (IL-6/BSF2) in Castelman's' disease. Blood 1989; 74:1360-67.

32. Beck JT, Hsu S, Wijdenes J et al. Brief report: Alleviation of systemic manifestations of Castelman's disease by monoclonal anti-interleukin-6 antibody. N Engl J Med 1994; 330:602-05.

INDEX

MOLECULAR BIOLOGY
INTELLIGENCE UNIT

AVAILABLE AND UPCOMING TITLES

NEUROSCIENCE INTELLIGENCE UNIT

AVAILABLE AND UPCOMING TITLES

MEDICAL INTELLIGENCE UNIT

AVAILABLE AND UPCOMING TITLES

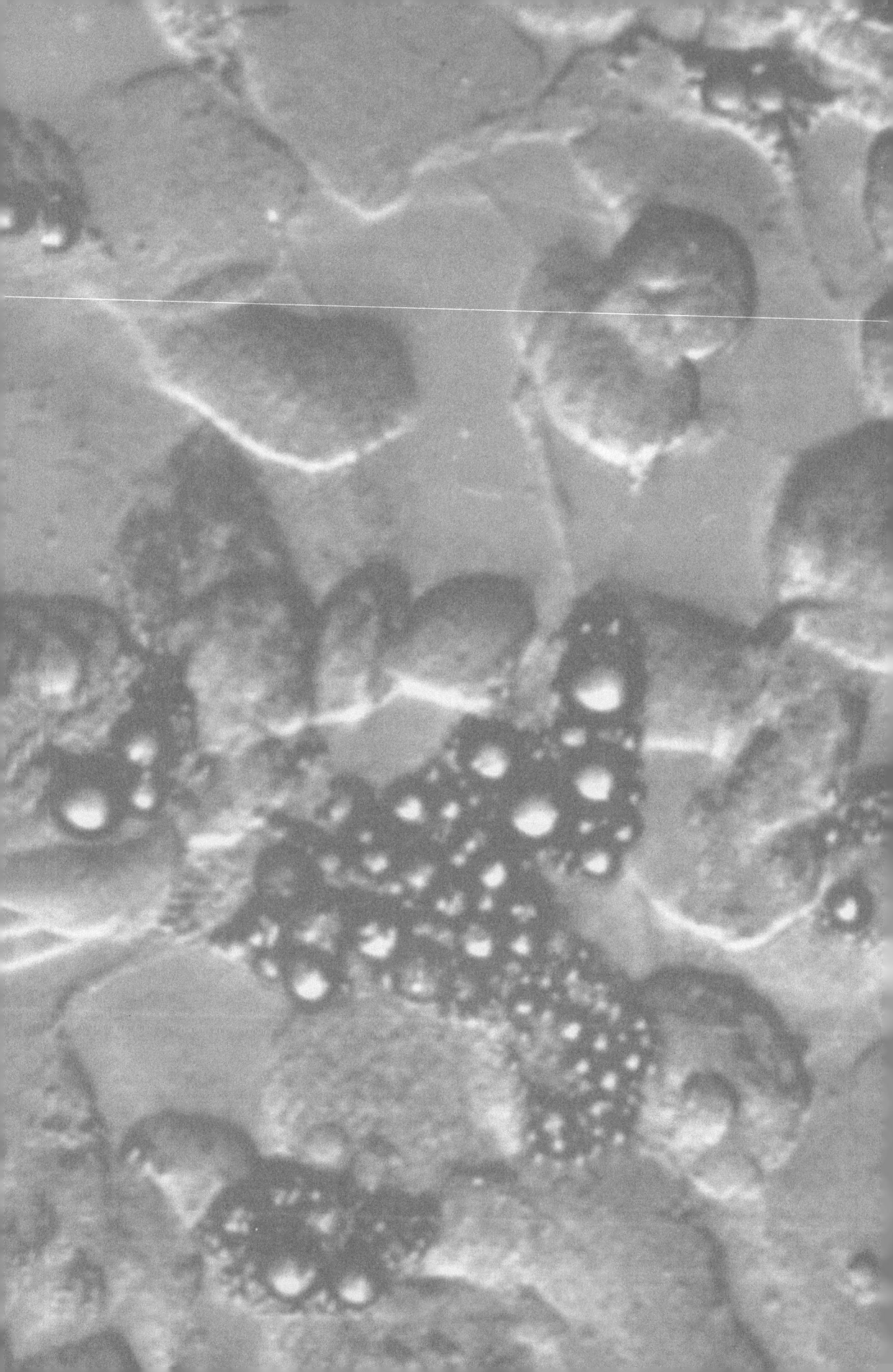

Effect of IL-10 and anti-TGF-beta
antibodies on the morphology of
bone marrow stroma cultures
from
Interleukin-10

Jan E. DeVries and
René de Waal Malefyt
© R.G. Landes Co. 1995